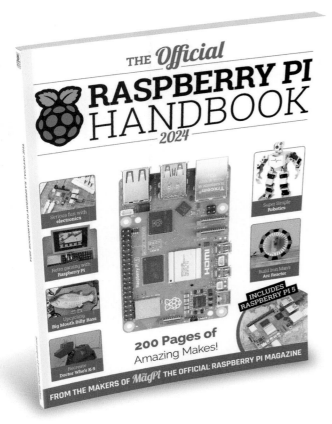

WELCOME!

Every year I have a real struggle choosing what projects and articles to put in the Official Raspberry Pi Handbook because, frankly, the community creates such vast quantities of incredible projects that cutting any seems like a disservice. This year was no different but looking over the list I was just flabbergasted at how incredible they all were. I always knew you could do some very cool stuff with Raspberry Pi but it's nice to have evidence to point to.

Elsewhere in the book we have detailed information on the brand spanking new Raspberry Pi 5, the best and fastest Raspberry Pi ever. Learn about the ten new components (including PCIe!) on the board and how they'll send your projects to the next level. We're all very excited about this one.

I'll leave you to read this book and find out about all the other incredible articles stuffed into these 200ish pages. Be excellent to each other and happy making.

Rob Zwetsloot

FIND US ONLINE magpi.cc | **GET IN TOUCH** magpi@raspberrypi.com

---

EDITORIAL
Editor: **Lucy Hattersley**
Features Editor: **Rob Zwetsloot**
Contributors: **PJ Evans, Rosie Hattersley, Phil King, Nicola King, Simon Monks, KG Orphanides, Toby Roberts**

DISTRIBUTION
Seymour Distribution Ltd
2 East Poultry Ave, London,
EC1A 9PT | **+44 (0)207 429 4000**

DESIGN
Critical Media: **criticalmedia.co.uk**
Head of Design: **Lee Allen**
Designers: **Olivia Mitchell, Sam Ribbits**
Illustrator: **Sam Alder**

MAGAZINE SUBSCRIPTIONS
Unit 6, The Enterprise Centre,
Kelvin Lane, Manor Royal,
Crawley, West Sussex,
RH10 9PE | **+44 (0)207 429 4000**
magpi.cc/subscribe
magpi@subscriptionhelpline.co.uk

PUBLISHING
Publishing Director: **Brian Jepson**
brian.jepson@raspberrypi.com

Advertising: **Charlotte Milligan**
charlotte.milligan@raspberrypi.com
Tel: +44 (0)7725 368887

Director of Communications: **Liz Upton**
CEO: **Eben Upton**

Contents

06

Introducing Raspberry Pi 5

14 Raspberry Pi Quickstart Guide
How to set up Raspberry Pi, from Zero to 400 and A to B+

Project Showcases

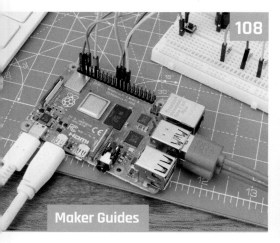

108

Maker Guides

154

Reviews and Resources

Introducing...

Raspberry Pi 5

Up to **three times faster**, absolutely packed with new features, and available to *The MagPi* subscribers first, Raspberry Pi 5 is everything we wanted from a new Raspberry Pi computer

By Lucy Hattersley

Raspberry Pi 5. Those are all the words you need to hear this month. This is the moment we've been waiting for.

The next generation of Raspberry Pi has been in development for years, and we can finally lift the wraps on the full specifications, design process and features on the new board.

For the first time, this computer features silicon designed in-house by Raspberry Pi. The new chip RP1 allows Raspberry Pi 5 to deliver a huge improvement in peripheral performance and functionality. It's also opened up a lot of space on the board, which now features 10 new components on the classic form factor.

Raspberry Pi 5 is still the single-board

computer we know and love, but at its heart sits a 2.4GHz quad-core Arm Cortex-A76 CPU that makes Raspberry Pi 5 between two and three times as fast as its predecessor.

Two models are available at launch: 4GB and 8GB. Both with SDRAM running at 4267 MHz.

The board remains the same size and dimensions but is packed with new features. It has a power button next to the status LED, a real-time clock, a Raspberry Pi connector for PCIe to support fast peripherals including NVMe drives (there is still an SD Card slot, now running twice as fast as before); there are two 4-lane DSI/CSI sockets, which support either two camera modules, two displays, or one of each. It has two HDMI connectors and now you can drive two full 4K displays both running at 60fps, up from 30fps on Raspberry Pi 4. There's even an UART connector, and a fan connector to provide power to a new case with an integrated fan.

Raspberry Pi 5 will start shipping in late October, and subscribers to The MagPi can pre-order theirs today with Priority Boarding. Raspberry Pi 5 is an excellent upgrade to our favourite micro-computer. This is the springboard we've been bouncing on for months, and we can't wait to dive in.

When is it out?

Raspberry Pi 5 will be available in 4GB and 8GB models from late October. Subscribers to *The MagPi* will get a pre-order code to buy Raspberry Pi 5 in the first wave.

Locate a reseller near you at the Raspberry Pi product page (**magpi.cc/products**)

10 New Components

Raspberry Pi's engineers have squeezed an incredible amount of extra tech onto the same-sized board

1 **BCM2712 Broadcom chip**
2 **RP1 I/O controller**
3 **Power button**
4 **Dual DSI/CSI sockets**
5 **Raspberry Pi connector for PCIe**
6 **UART debug connector**
7 **Fan & Active Cooler power connector**
8 **Real-time clock with battery connector**
9 **Power-management IC**
10 **Heatsink mounts**

Get to know
Raspberry Pi 5

Your in-depth guide to Raspberry Pi's new micro-computer

Specifications

2.4GHz 64-bit quad-core Cortex-A76 processor

VideoCore VII GPU

4GB / 8GB LPDDR4X SDRAM at 4267MHz

microSD (SDR104 supported)

2 × micro HDMI ports (supports up to 4Kp60)

2 × USB 3.0 ports

2 × USB 2.0 ports

2 × 4-lane connectors for camera or display peripherals (sold separately)

Gigabit Ethernet port

802.11b/g/n/ac wireless

Bluetooth 5.0

PoE-capable (requires PoE HAT, sold separately)

Raspberry Pi connector for PCIe (requires M.2 HAT, sold separately)

Power button

Fan connector

UART connector

Real-time clock, with connector for battery backup

5V/5A USB-C power supply recommended, 5V/3A minimum requirement (sold separately)

RAM
Here we have 8GB of LPDDR4 in the form of a Micron chip

PCIe
A new PCI Express high-speed expansion bus is positioned on the edge of the board. An upcoming adaptor will enable an M.2 drive to be connected directly to Raspberry Pi 5, along with other custom devices.

Power
The Renesas/Dialog DA9091 "Gilmour" power management chip is custom silicon that provides power supplied to the various components

Power Button
A power button, one of the most heavily requested features, has finally been added to Raspberry Pi 5. The status LED sits adjacent

USB-C power
Raspberry Pi 5 is powered by an external USB-C power supply

BCM2712 chip

The silver heart of Raspberry Pi 5 is the new Broadcom BCM2712 SoC (System-on-chip) architecture. It contains an ARM Cortex-A76 quad-core CPU running at 2.4GHz and a new VideoCore VII GPU that supports OpenGL-ES 3.1, Vulkan 1.2.

Fan socket

A new on-board fan socket is used to provide power to the fan-enabled case and Active Cooler accessories

Active Cooler

There are two extra holes on the board. These are to mount the Active Cooler attachment

Raspberry Pi RP1

The new Raspberry Pi RP1 chip handles the bulk of input and output (I/O). It is connected to BCM2712 via PCI Express.

Dual 4Kp60 HDMI

There are two micro-HDMI ports; Raspberry Pi 5 can drive dual displays, both at 4Kp60 resolution

Dual CSI/DSI

The CSI and DSI ports have been combined into two multipurpose CSI/DSI ports (now using the denser connector pinout found on Raspberry Pi Zero). You can connect either two displays, or two cameras (or one of each)

UART

Between the micro-HDMI ports sits a new UART connector, which can be used to control Raspberry Pi 5 in headless mode

RTC BATTERY

The battery connector is used to connect a battery (or supercapacitor) to provide backup power to the Real Time Clock

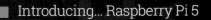

▶ On the reverse of Raspberry Pi 5 we can see the SD card slot and connectors soldered together in more detail

Battery

Raspberry Pi 5 has a two-pin JST (Japan Solderless Terminal) port marked BAT. This is used to connect a battery (or other power source) to the new Real Time Clock (RTC).

Raspberry Pi RP1

Raspberry Pi has been developing RP1 for a long time. RP1 is a new communications chip. RP1 is custom silicon designed by Raspberry Pi that connects the CPU to the 'slower' components on the board (of course, these slow components can still function incredibly quickly).

In the case of Raspberry Pi 5, RP1 controls the input and output (I/O) for the GPIO pins, USB ports, CSI/DSI ports and Ethernet. It is connected to the BCM2712 SoC via a 4-lane PCI-express bus. The GPIO pins have the same function and layout as before, so most HATs and other accessories will be compatible.

Explore the new
Raspberry Pi OS

A new operating system is also announced: Raspberry Pi OS based on Debian 'bookworm'

Raspberry Pi 5 can run a variety of different operating systems and, thanks to its speed increase, you have a lot of Linux and other ARM-based OSes to choose from.

Raspberry Pi OS is still our preferred operating system. Built on top of Debian, it is a Linux operating system custom-built for Raspberry Pi hardware.

With Raspberry Pi OS you get easy access to hardware features like the GPIO pins, and Camera Module, and can be sure everything is being tested. Out shortly before Raspberry Pi 5 is available will be a new version of Raspberry Pi OS, based on Debian 'bookworm'.

Raspberry Pi 4 and Raspberry Pi 5 hardware running bookworm replaces the X11-based LXDE desktop with a Wayland system using Wayfire as the window manager. This software is used to draw windows on the screen and is highly customizable. Raspberry Pi OS has a slicker windowing system with subtle animations. The desktop background is still drawn by pcmanfm but has been modified to talk to Wayland. The taskbar looks the same but is a customized version of Wayfire's wf-panel, which also includes ports of the existing plugins.

The experience of Raspberry Pi OS (bookworm)

▼ Firefox joins Chrome as a web browser option

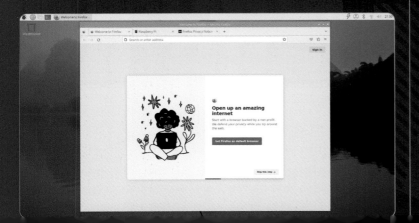

on Raspberry Pi 5 is fantastic, offering similar levels of interactive speed as a desktop-class computer. It has the same feature set and design as before, but with vastly improved performance and a slightly slicker design: windows fade in and out with a smooth animation.

There are new desktop images as well, but the overall style is in keeping with earlier versions of Raspberry Pi OS.

There is also a new Network Manager (**networkmanager.dev**) which replaces dhcpcd. Network Manager is becoming the standard networking tool on many Linux distributions, while dhcpcd was relatively uncommon.

> ❝ The experience of Raspberry Pi OS (bookworm) on Raspberry Pi 5 is fantastic ❞

Pipewire is now being used instead of PulseAudio to handle audio and video. This is used by some tools to provide services like screen sharing for Wayland, so it is a useful enabling technology.

Finally, Firefox joins Google Chrome as a browser option in the Recommended installation of Raspberry Pi OS. Firefox enables browser sync across Raspberry Pi and other operating systems, a feature missing from Chrome.

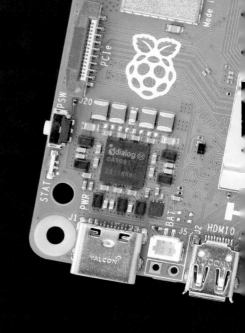

Raspberry Pi 5
Accessories

Modular design is more feature-packed than ever

Raspberry Pi Case

Alongside Raspberry Pi 5 comes a new case with powerful new features.

It retains the white and red modular scheme of the original case, and now contains four separate compartments.

There is the red base, which Raspberry Pi 5 sits on, with a white frame in the middle. New to this case is a clear shelf with a fan on board (that connects to the fan socket next to the Ethernet port). This shelf has a cutout section enabling access to the GPIO pins, and a final lid, with airflow spacing, covers everything into a neat package. Inside the box you will also find a heatsink to attach to Raspberry Pi 5's main processor and four stick-on rubber feet.

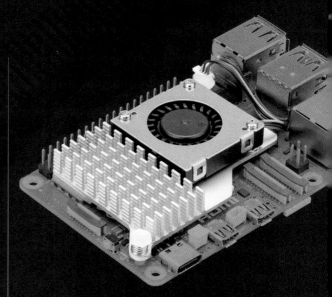

Active Cooler

An alternative cooling solution is also available, called the Active Cooler. It may look like a fan, but this is an extruded heatsink with fins and a blower (a fan that sucks air in from the top and directs it sideways). It connects directly to the board for power. You'll notice two extra holes on Raspberry Pi 5's board to attach the Active Cooler, which is connected with pushpins.

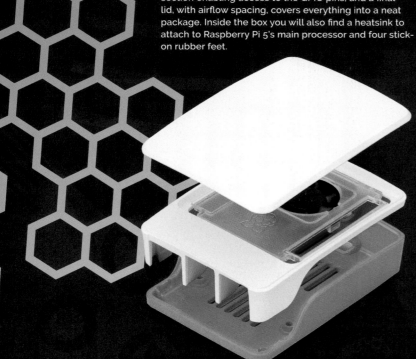

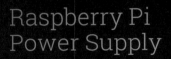

Raspberry Pi Power Supply

Raspberry Pi 5 requires a 25 W (5 V/5 A) USB-C power supply to support maximum current to downstream USB devices. With only low-power peripherals like mice and keyboards attached, you can safely continue to use a 5 V/3 A (15 W) PSU as a minimum requirement. In testing, we have been using the 15 W power supply without the Active Cooler or Raspberry Pi Case without issues.

To power Raspberry Pi 5 to the best of its abilities, you should invest in the new Raspberry Pi Power Supply, which provides 25 W of power via a USB-C connection. The Power Supply's input voltage is 100–240 V~50/60 Hz, and output modes are:

- **5.1 V/5 A (25.5 W)**
- 9 V/3 A (27 W)
- **12 V/2.25 A (27 W)**
- 15 V/1.8 A (27 W)

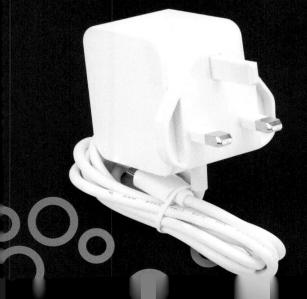

PCI Express M.2 adapter

We can exclusively reveal that Raspberry Pi is working on a PCI Express adapter to connect M.2 storage drives directly to Raspberry Pi 5. We'll have more information on this soon because the L-shaped design is being finalised. In the meantime, here is a photo of the adapter cable that will connect to the M.2 HAT. 🅜

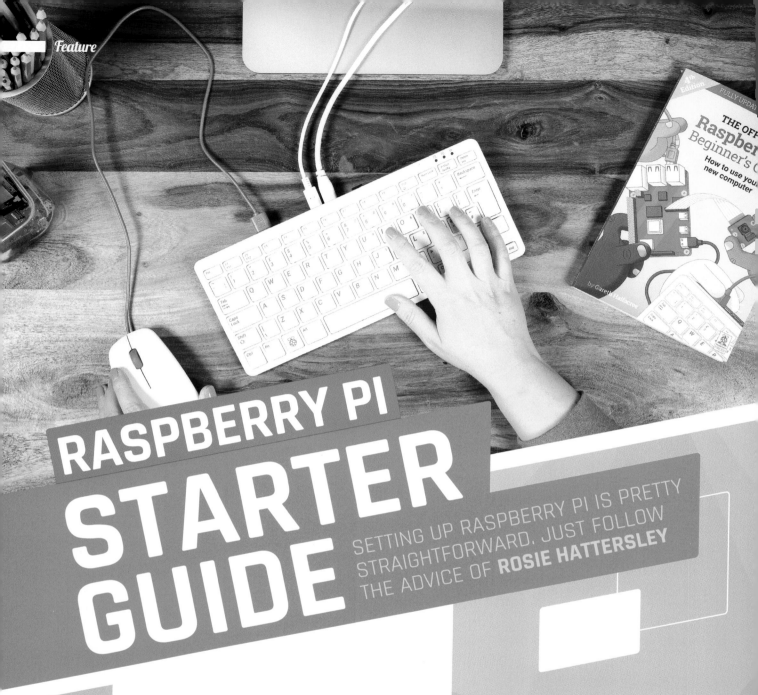

RASPBERRY PI
STARTER GUIDE

SETTING UP RASPBERRY PI IS PRETTY STRAIGHTFORWARD. JUST FOLLOW THE ADVICE OF **ROSIE HATTERSLEY**

Congratulations on becoming a Raspberry Pi explorer. We're sure you'll enjoy discovering a whole new world of computing and the chance to handcraft your own games, control your own robots and machines, and share your experiences with other Raspberry Pi fans.

Getting started won't take long: just corral the extra bits and bobs you need on our checklist. Useful additions include some headphones or speakers if you're keen on using Raspberry Pi as a media centre, or a gamepad for use as a retro games console.

To get set up, simply use your pre-written microSD card (or use Raspberry Pi Imager to set up a card) and connect all the cables. This guide will lead you through each step. You'll find the Raspberry Pi OS, including coding programs and office software, all available to use. After that, the world of digital making with Raspberry Pi awaits you.

What you need

All the bits and bobs you need to set up a Raspberry Pi computer

A Raspberry Pi

Whether you choose the new Raspberry Pi 5; or a Raspberry Pi 4, 400, 3B+, 3B; Raspberry Pi Zero or Zero 2 (or an older model of Raspberry Pi), basic setup is the same. All Raspberry Pi computers run from a microSD card, use a USB power supply, and feature the same operating systems, programs, and games.

8GB microSD card

You'll need a microSD card with a capacity of 8GB or greater. Your Raspberry Pi uses it to store games, programs, and boot the operating system. Many Raspberry Pi computer kits come with a card pre-written with Raspberry Pi OS. If you want to reuse an old card, you'll need a card reader: either USB or a microSD to full-sized SD (pictured).

Windows/Linux PC or Mac computer

You'll need a computer to write Raspberry Pi OS to the microSD card. It doesn't matter what operating system this computer runs, because it's just for installing the OS using the Raspberry Pi Imager app.

USB keyboard

Like any computer, you need a means to enter web addresses, type commands, and otherwise control Raspberry Pi. The Raspberry Pi 400 comes with its own keyboard. Raspberry Pi sells an official Keyboard and Hub (**magpi.cc/keyboard**) for other models.

USB mouse

A tethered mouse that physically attaches to your Raspberry Pi via a USB port is simplest and, unlike a Bluetooth version, is less likely to get lost just when you need it. Like the keyboard, we think it's best to perform the setup with a wired mouse. Raspberry Pi sells an Official Mouse (**magpi.cc/mouse**).

Power supply

Raspberry Pi 4 and Raspberry Pi 400 need a USB Type-C power supply. Raspberry Pi sells power supplies (**magpi.cc/usbcpower**), which provide a reliable source of power. Raspberry Pi 1, 2, 3, and Zero models need a micro USB power supply (**magpi.cc/universalpower**).

Display and HDMI cable

A standard PC monitor is ideal, as the screen will be large enough to read comfortably. It needs to have an HDMI connection, as that's what's fitted on your Raspberry Pi board. Raspberry Pi 4 and 400 can power two HDMI displays, but require a micro-HDMI to HDMI cable. Raspberry Pi 3B+ and 3A+ both use regular HDMI cables; Raspberry Pi Zero W needs a mini-HDMI to HDMI cable (or adapter).

USB hub

Raspberry Pi Zero and Model A boards have a single USB socket. To attach a keyboard and mouse (and other items), you should get a four-port USB hub (or use the official USB Keyboard and Hub which includes three ports). Instead of standard-size USB ports, Raspberry Pi Zero has a micro USB port (and usually comes with a micro USB to USB-A adapter).

SET UP RASPBERRY PI

Raspberry Pi 5 / 4 / 3B+ / 3 has plenty of connections, making it easy to set up

01 Hook up the keyboard

Connect a regular wired PC (or Mac) keyboard to one of the four larger USB-A sockets on a Raspberry Pi 5 / 4 / 3/3B+. It doesn't matter which USB-A socket you connect it to. It is possible to connect a Bluetooth keyboard, but it's much better to use a wired keyboard to start with.

02 Connect a mouse

Connect a USB wired mouse to one of the other larger USB-A sockets on Raspberry Pi. As with the keyboard, it is possible to use a Bluetooth wireless mouse, but setup is much easier with a wired connection.

03 HDMI cable

Next, connect Raspberry Pi to your display using an HDMI cable. This will connect to one of the micro-HDMI sockets on the side of a Raspberry Pi 5 / 4, or full-size HDMI socket on a Raspberry Pi 3/3B+. Connect the other end of the HDMI cable to an HDMI monitor or television.

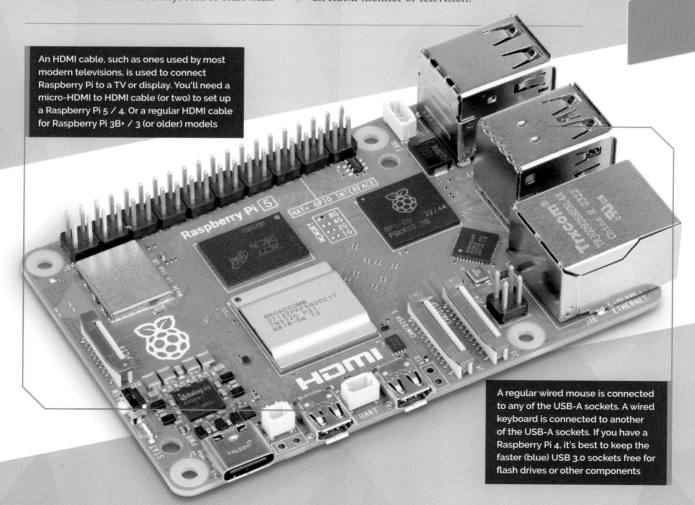

An HDMI cable, such as ones used by most modern televisions, is used to connect Raspberry Pi to a TV or display. You'll need a micro-HDMI to HDMI cable (or two) to set up a Raspberry Pi 5 / 4. Or a regular HDMI cable for Raspberry Pi 3B+ / 3 (or older) models

A regular wired mouse is connected to any of the USB-A sockets. A wired keyboard is connected to another of the USB-A sockets. If you have a Raspberry Pi 4, it's best to keep the faster (blue) USB 3.0 sockets free for flash drives or other components

The USB-C socket is used to connect power to the Raspberry Pi 400. You can use a compatible USB-C power adapter (found on recent mobile phones) or use a bespoke power adapter such as the Raspberry Pi 15.3 W USB-C Power Supply

The Ethernet socket can be used to connect Raspberry Pi 400 directly to a network router (such as a modem/router at home) and get internet access. Alternatively, you can choose a wireless LAN network during the welcome process

SET UP RASPBERRY PI 400

Raspberry Pi 400 has its own keyboard – all you need to connect is the mouse and power

01 Connect a mouse
Connect a wired USB mouse to the white USB connection on the rear of Raspberry Pi 400. The two blue USB 3.0 connectors are faster and best reserved for external drives and other equipment that need the speed.

02 Attach the micro-HDMI cable
Next, connect a micro-HDMI cable to one of the micro-HDMI sockets on the rear of Raspberry Pi 400. The one next to the microSD card slot is the first one, but either connection should work. Connect the other end of the HDMI cable to an HDMI monitor or television.

03 The microSD
If you purchased a Raspberry Pi 400 Personal Computer Kit, the microSD card will come with Raspberry Pi OS pre-installed. All you need to do is connect the power and follow the welcome instructions. If you do not have a Raspberry Pi OS pre-installed microSD card, follow the instructions later in 'Set up the software'.

You'll need this micro USB to USB-A adapter to connect wired USB devices such as a mouse and keyboard to your Raspberry Pi Zero W

Raspberry Pi Zero W features a mini-HDMI socket. You'll need a mini-HDMI to full-sized HDMI adapter like this to connect your Raspberry Pi Zero W to an HDMI display

SET UP RASPBERRY PI ZERO

You'll need a couple of adapters to set up Raspberry Pi Zero and Zero 2

01 Get it connected

If you're setting up a smaller Raspberry Pi Zero, you'll need to use a micro USB to USB-A adapter cable to connect the keyboard to the smaller connection on the board. Raspberry Pi Zero models only have a single micro USB port for connecting devices, which means you'll need to either get a small USB hub or use an all-in-one mouse and keyboard.

02 Mouse and keyboard

You can either connect your mouse to a USB socket on your keyboard (if one is available), then connect the keyboard to the micro USB socket (via the micro USB to USB-A adapter). Or, you can attach a USB hub to the micro USB to USB-A adapter.

03 More connections

Now connect your full-sized HDMI cable to the mini-HDMI to HDMI adapter, and plug the adapter into the mini-HDMI port in the middle of your Raspberry Pi Zero. Connect the other end of the HDMI cable to an HDMI monitor or television.

First, insert your microSD card into Raspberry Pi

With the microSD card fully inserted, connect your power supply cable to Raspberry Pi. A red light will appear on the board to indicate the presence of power

SET UP
THE SOFTWARE

Use Imager to install Raspberry Pi OS on your microSD card and start your Raspberry Pi

Now you've got all the pieces together, it's time to install an operating system on your Raspberry Pi so you can start using it. Raspberry Pi OS is the official software for Raspberry Pi, and the easiest way to set it up on your Raspberry Pi is to use Raspberry Pi Imager. See the 'You'll Need' box and get your kit together.

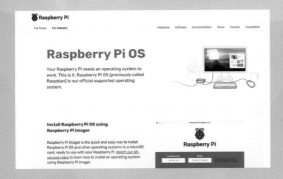

01 Download Imager
Raspberry Pi Imager is available for Windows, Mac, and Linux computers. You can also install it on Raspberry Pi computers, to make more microSD cards once you are up-and-running. Open a web browser on your computer and visit **magpi.cc/imager**. Once installed, open Imager and plug your microSD card into your computer.

02 Choose OS
Click on 'Choose OS' in Raspberry Pi Imager and select the recommended Raspberry Pi OS. Click 'Choose SD card' and select the microSD card you just inserted (it should say 8GB or the size of the card next to it). Click on 'Write'. Your computer will take a few minutes to download the Raspberry Pi OS files, copy them to the microSD card, and verify that the data has been copied correctly.

03 Set up Raspberry Pi
When Raspberry Pi Imager has finished verifying the software, you will get a notification window. Remove the microSD card and put it in your Raspberry Pi. Plug in your Raspberry Pi power supply and, after a few seconds, a blue screen will appear with 'Resizing Filesystem'. It will quickly vanish and be replaced by 'Welcome to Raspberry Pi'. Click on Next and follow the on-screen instructions to set up Raspberry Pi OS and start using your new computer.

Top Tip

Is your card ready?

You don't need to do this if your Raspberry Pi came with a card pre-written with Raspberry Pi OS.

You'll Need

- A Windows/Linux PC or Apple Mac computer
- A microSD card (8GB or larger)
- A microSD to USB adapter (or a microSD to SD adapter and SD card slot on your computer)
- Raspberry Pi Imager **magpi.cc/imager**

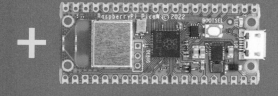

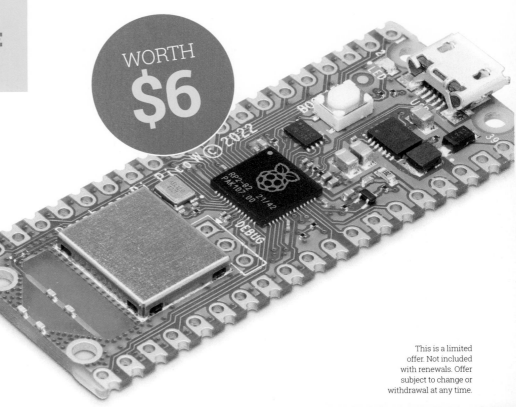

Project Showcases

86

102

70

PiGlass V2

A new Raspberry Pi Zero means a new take on PiGlass, the smart glasses powered by Raspberry Pi. **Rob Zwetsloot** takes a look

MAKER

Matt Desmaris

Matt is a volunteer at his local food pantry as the IT department, and has been using Raspberry Pi for ten years.

mrdcreations.org

Do you remember Google Glass? They were smart glasses with a bit of augmented reality thrown in. Unfortunately, they never became a full product; however, with advancements in computer vision and miniaturisation of computers, custom smart glass builds never really went away.

One maker, Matt Desmaris, recently revisited the idea. "I made PiGlass V1 in 2018," he explains. "I wanted to try to make a heads-up display wearable and see how far I could take it. I kept running into performance issues with Raspberry Pi Zero when I was trying to add more features, and I made a note to revisit the project when Raspberry Pi Zero 2 came out."

PiGlass V2 has some extra features, like bone conduction earphones. It also makes use of a wearer's hat.

▶ A better view of the wiring of the system

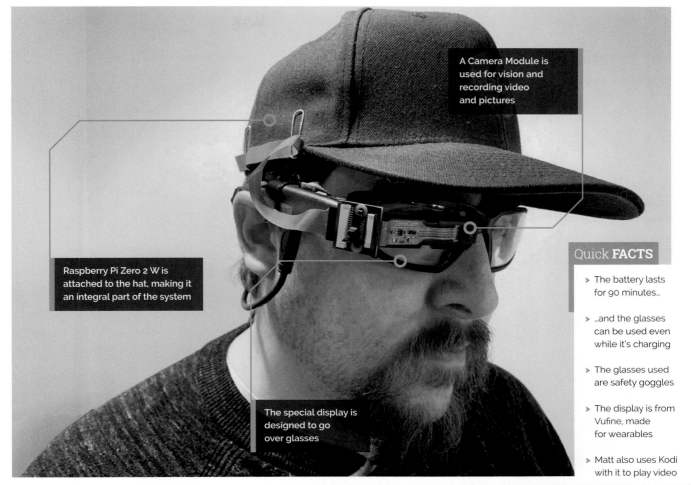

A Camera Module is used for vision and recording video and pictures

Raspberry Pi Zero 2 W is attached to the hat, making it an integral part of the system

The special display is designed to go over glasses

Quick FACTS

> The battery lasts for 90 minutes...

> ...and the glasses can be used even while it's charging

> The glasses used are safety goggles

> The display is from Vufine, made for wearables

> Matt also uses Kodi with it to play video

"The whole system is controlled with a gamepad, which makes for an interesting sight"

Virtual vision

Matt was in contact with us throughout the process of making this and the build, actually, was quite quick, as he elaborates: "Construction was straightforward – a soldering iron, heat gun, and small flat-head [screwdriver] are all that's required. Construction took place over a couple weeks as I progressively got more and more parts in. Everything is secured with zip ties or heat shrink."

The whole system is controlled with a gamepad, which makes for an interesting sight.

"There is a button on the audio hat on the back of my head," Matt reveals. "It is the start/stop button. Start the menu program or

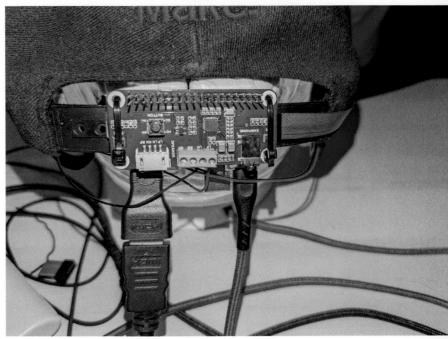

▲ The main system is mounted on the rear of the hat, feeding to the display

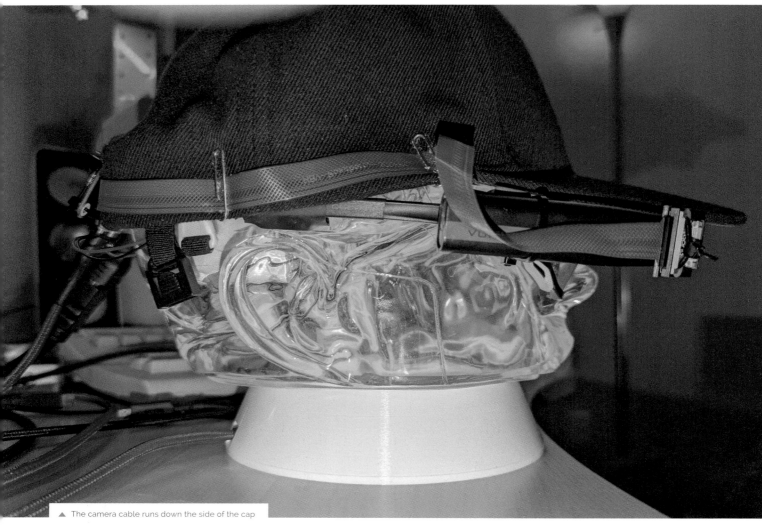

▲ The camera cable runs down the side of the cap

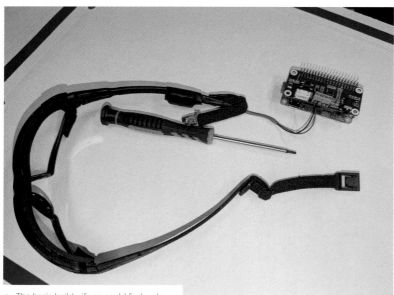

▲ The basic build - if you could find a place for Zero 2 W. you don't need a hat

kill every program that could be running. The menu program uses the [picamera] API which allows it to be recorded, including the text overlays. The menu program has a few options: camera (from PiGlass V1) at 1080p, record video with audio at 1080p, stream YouTube at 720p, Emulation Station, Kodi at 720p, and Steam Link [with controller issues]."

While Matt developed it hooked up to a monitor, he's tested it thoroughly by watching streaming video and playing some retro games – something we heartily approve of.

Seeing cyber

"Within the menu/related programs it works very well; some things take a few seconds to load," Matt tells us. "The 720p display looks really good; all the text in menus and captions are easily readable. The camera program allows you to take 1080p images/silent videos, and video with audio records at 1080p, 25fps, with 44kB audio.

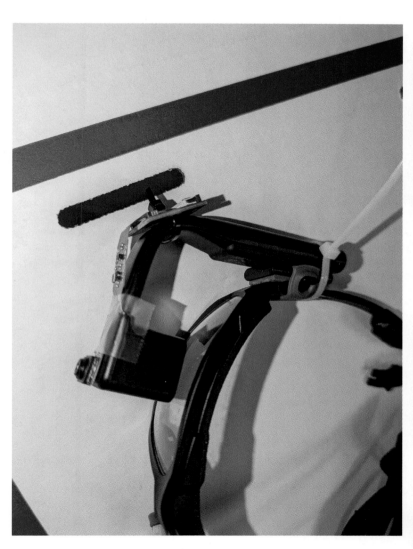

The display is designed for this kind of use

Using smart glasses

01 Using the controller, you can choose what function you'd like to use. These are displayed as words over the feed image.

02 Watching streaming services will allow you to sit back and relax on a commute, or while you wait for something.

03 For gaming, you're able to use the controller you use for the rest of navigation, making the whole process more streamlined.

YouTube stream will livestream to YouTube in 720p, 25 fps, with 44kB audio."

With the project now at an advanced stage, Matt has been thinking of future improvements as well: "I want to add voice commands and I have a few ideas. The microphones are located on the back of my head and they can pick up my voice at normal speaking levels. I want to add real-time object detection. I have tested demo code and it looks like [Raspberry Pi] Zero 2 W can do it. I want to make a program that uses the gamepad to be able to select which type of object is being detected."

We very much look forward to the cyberpunk cyberglass future. 🅼

Fancy Octopus Arcades

Inspiring the next generation of gamers led to a business creating custom arcade housings, discovers **Rosie Hattersley**

MAKER

Shonee Strother

Shonee swapped a role as creative director and fulfilled a childhood dream to build gaming arcades, with assistance from his "highly talented" six-year-old son.

magpi.cc/ fancyoctopus

"**W**e're in the golden age of emulation," declares Shonee Strother.** He doesn't have a standard technical background, but he's been a gamer all his life and, when the pandemic put his role as a creative director on hold, he began looking for new hobbies. Shonee and his six-year-old son Wolf started their Raspberry Pi adventures "by modding an arcade 1up with a Raspberry Pi 3B." Wolf is extremely interested in gaming, like his dad, who felt that his son's young age meant "there's so much gaming history he's missed. Giving him a slice of that was super important." When Shonee documented their work on Instagram, people began to take notice. Their bespoke gaming arcade business, Fancy Octopus Arcades, "just sort of took off from there."

Bijou is beautiful

Brooklyn resident Shonee started following a few Reddit and Facebook pages where he "saw folks making huge gaming rooms, loading them up with

▶ The unique retro arcade builds have proved extremely popular, with handheld designs planned soon

arcade cabinets, and realised that for some of us in smaller apartments, that wasn't a possibility." Shonee was also frustrated that people were capitalising on the scarcity of arcade cabinets, "snagging all the 1up machines they could find and price gouging them to death." He decided to address pent-up demand for games arcades, focusing on designs that would fit a New York-sized apartment.

The idea was to make small, mini cabinets, packed to the gills with games, easily plugged via standard HDMI for video audio output. Shonee also wanted them to be highly personalised art pieces "that a person could be proud to put on a display shelf or coffee table." Raspberry Pi's small profile and extremely robust performance helped him "pack a lot of oomph into a very small package," he confirms. He also loves being able to easily flash images directly onto the SD card with little or no manual coding changes. As a result, he can focus on each cabinet's artistry.

Daring designs

Although homemade retro games cabinets are a popular build, Shonee's stand out because he designs and creates all the parts himself. The 3D printing, silkscreen, vinyl wrap, and decoupage work is all done in-house, depending on the build. Shonee works closely with clients, creating a design around their vision. "No two builds are ever alike, and no two designs are ever repeated," he says of what is now a two-year-old business. "I've had so many folks ask me 'can you build me a copy of that Donkey Kong deck?' My bank account absolutely hates that I have to say no." The unique elements that bring his designs to life include gear sticks masquerading as swords for his Samurai

Although each build is different visually, inside there's a Raspberry Pi running RetroPie games emulators and ArcadePunks' front end software

The portable arcade cabinets use zero delay encoders wired to 28 mm LED arcade buttons and also feature some powerful internal speakers

'The custom-designed exteriors involve skills such as vinyl cutting, painted artwork, 3D-printed elements, and unexpected extra features

Quick **FACTS**

> Each arcade Shonee builds has a secret, unannounced feature

> The Zelda one has a rainbow LED rupee-filled chest on top

> It's a music box that plays the Zelda fairy theme

> The arcade can be updated remotely and run off rechargeable batteries...

> ...Ideal for participating in a recent MVC2 battle in Central Park

▲ Arcade cabinet designs feature elements from the client's favourite games

▲ Following the success of the bespoke arcades, Shonee is keen to branch out into custom handheld consoles

◀ This RetroPie arcade features a barrel-shaped planter from a garden store

▲ The handle for this Haring vs Basquiat build is from a 1980s Panasonic boom box that Shonee bought at a NY flea market

Pimp your arcade

Build the arcade cabinet, either by downloading and printing designs from a site like Thingiverse or creating a custom wood build. You'll need a Raspberry Pi, zero delay encoders for buttons and joysticks, speakers, a mini amp, and some creativity.

01 Wire up the unit

Connect zero delay encoders to 28 mm LED arcade buttons. For internal audio, add a mini amp and stereo speakers mounted to the control deck.

02 Allow for updates

Connect Raspberry Pi to an external Ethernet port for any updates or system changes. Install a USB hub to power the LEDs, and attach a mini keyboard for any edits that may be needed.

03 Testing, testing

Flash-update RetroPie and run it to make sure the front end works with the library you've installed. Depending on the display you're using, check for scan lines, and adjust frame rates and the aspect ratio.

Jack build, the handle from a Panasonic boom box found at a Chinatown flea market and used in his Basquiat/Keith Haring build, and a Donkey Kong barrel refashioned from a garden centre planter. "Basically, I use whatever I can." More often than not, he uses software from **Arcadepunks.com**. Their front ends are "stable, responsive, and rarely cause problems down the line," Shonee says.

"Some folks really want to focus on retro arcade gaming; some folks are more focused on having a game preservation library of console games from their past. There are always some adjustments that need to be made. Luckily, with the versatility of Raspberry Pi, it's pretty easy." Each Fancy Octopus Arcade typically takes about a month to complete, with costs varying depending on the brief.

"Without Raspberry Pi, this would just be a box with buttons on it," Shonee concludes. "The size, power, versatility, and ease of use have given me the ability to help the dream of making my hobby a career come to fruition." 🎮

ZX Spectrum
Raspberry Pi Cassette

Between jobs, one maker decided to push their Raspberry Pi skills and
make a portable ZX Spectrum Raspberry Pi. **Rosie Hattersley** approves

MAKER

Stuart Brand

Stuart aka
JamHamster is "an IT
Root Cause Engineer
for work and an avid
(some would say
obsessed) tinkerer in
my off hours!"

@RealJamhamster

Stuart Brand was between jobs and decided to concentrate on pushing his skills by building Raspberry Pi projects: **"I headed to the garage and embraced my inner nerd!"** exclaims the maker of the ZX Spectrum Raspberry Pi Cassette. "I wouldn't have had a clue how to build any of this stuff before lockdown. It goes to prove that you never know what you're capable of until you give it a go."

Stuart's first computer, a ZX Spectrum, has a special place in his heart, so a Raspberry Pi project based around one seemed ideal. "They're still great machines!" he says of the beloved computer which celebrates its 40th birthday this April.

Stuart loves repairing and running real hardware as well as emulations and thought "it would be

nifty to see if I could fit an entire ZX Spectrum emulator into a cassette tape shell." He now uses his ZX Spectrum Pi Cassette as a 'pick up and play' device whenever he fancies "a quick bash at some old school gaming."

Learn as you go

Prior to this project, Stuart had several retro makes under his belt and had made a tape emulator for an Arduino-based ZX Spectrum +2 that acts like a multi-cart tape. "Putting a whole Spectrum in a tape shell was the next logical step and an interesting challenge," he says. Being tight for space, he chose Raspberry Pi Zero W. He loves the fact both ZX Spectrum and Raspberry Pi's ARM processor were developed in Cambridge.

> ❝ Stuart wrapped Raspberry Pi in foil and went at it with a Dremel ❞

Despite this, he describes himself as a haphazard tinkerer with little electronics experience, who plans everything in his head. "I don't have any schematics to share," he apologises, "and never measure anything." However, he makes paper mock-ups of everything he's planning, largely to ensure it all fits. A veteran of small case builds, Stuart cautions other wannabe makers to leave far more room for cables than you think you'll need. He also admits to treating his Raspberry Pi collection rather roughly: "even though they have been abused and tortured, they still keep running."

Stuart assembled the ZX Spectrum Raspberry Pi build from what he had to hand. He took a sheet of scrap metal and used a bandsaw to fashion a crude

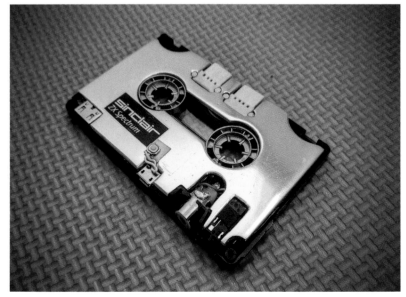

▲ The handcrafted heatsink fits beautifully inside the repurposed cassette tape

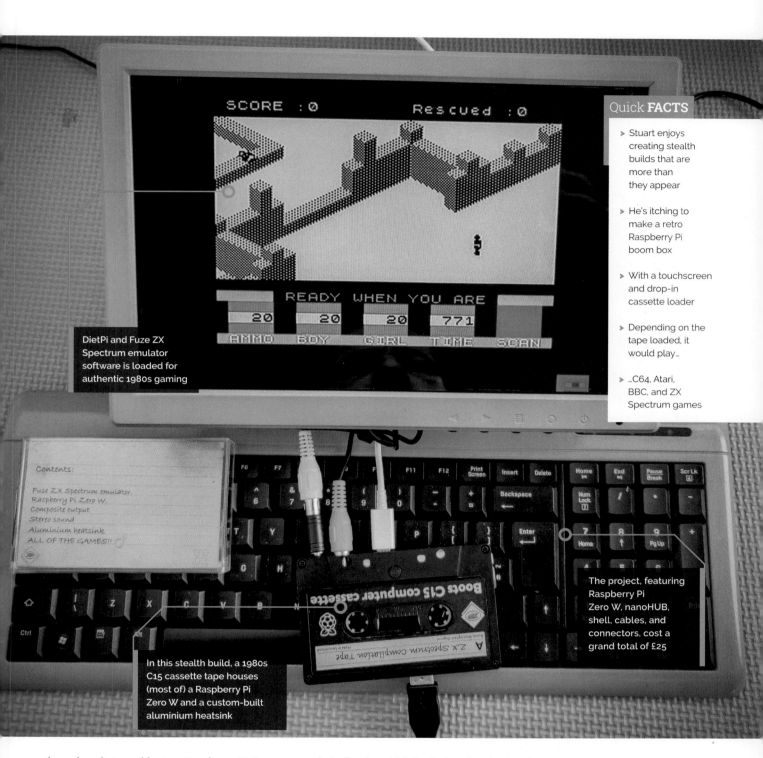

SCORE :0   Rescued :0

READY WHEN YOU ARE

| 20 | 20 | 20 | 771 | |
| AMMO | BOY | GIRL | TIME | SCAN |

DietPi and Fuze ZX Spectrum emulator software is loaded for authentic 1980s gaming

Contents:

Fuze ZX Spectrum emulator.
Raspberry Pi Zero W.
Composite output
Stereo sound
Aluminium heatsink
ALL OF THE GAMES!!

In this stealth build, a 1980s C15 cassette tape houses (most of) a Raspberry Pi Zero W and a custom-built aluminium heatsink

The project, featuring Raspberry Pi Zero W, nanoHUB, shell, cables, and connectors, cost a grand total of £25

Quick FACTS

> Stuart enjoys creating stealth builds that are more than they appear

> He's itching to make a retro Raspberry Pi boom box

> With a touchscreen and drop-in cassette loader

> Depending on the tape loaded, it would play...

> ...C64, Atari, BBC, and ZX Spectrum games

---

shape for what would act as Raspberry Pi Zero W's heatsink. A Dremel, needle files, and fine-grit sandpaper were used to finesse the shape.

Getting it taped

Stuart bought a job lot of cassette tape seconds: "Boots C15 were the cassettes I used for storing my programs back in the '80s; it was an obvious choice" – for which he designed and printed new labels. "Cassette shells make for a great form factor," says Stuart, "I started with a plain black spare cassette shell and used a small hand file and side cutters to remove the plastic supports in preparation for fitting the heatsink."

The 5 mm interior of the C15 cassette tape meant something would have to give: fitting a Zero W

▲ Stuart has form making retro gaming builds using Raspberry Pi

▶ Customised cassette labels complete the 1980s look

Alert!
Warranty voiding

This project involves cutting off the GPIO pins in order to fit Raspberry Pi Zero W inside a cassette tape. This risks damaging Pi Zero W and invalidates your warranty. Wrapping it in aluminium foil while cutting is a sensible precaution. Cut carefully.

> ▲ Stuart's cassette edition of the ZX Spectrum atop an original Spectrum keyboard

❝ I lost some GPIO ports, but it was well worth it to get the tape looking right ❞

inside involved cutting out a section to nestle under the reels and "preserve the illusion" – not something inexperienced makers are advised to tackle. Stuart has eight similar builds under his belt, hence his confidence. He wrapped Raspberry Pi in foil and "went at it with a Dremel." Surprisingly, it survived. "I lost some GPIO ports, but it was well worth it to get the tape looking right."

Configuring the DietPi and Fuze ZX Spectrum emulator took lots of tweaks before Stuart was able to get them to boot in an acceptable time frame. "I eventually got it to boot in 16 seconds. The full-width heatsink meant I could safely overclock Zero W and saved another couple of seconds," he says.

His next challenge: a 1980s boom box with drop-in cassettes that boot up and play games from different iconic home computers. We like his thinking! ◪

Make a tape

01 Install DietPi (**dietpi.com**) and the Fuze ZX Spectrum emulator (**fuse-emulator.sourceforge.net**) on your Raspberry Pi and set it to autorun at startup.

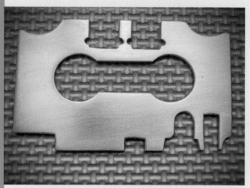

02 Create a heatsink following the contours of the cassette case, and avoiding the spool area. Wrap Raspberry Pi Zero W in aluminium foil and very carefully cut off the GPIO port section where it will prevent the spool wheels fitting.

03 Position Raspberry Pi Zero W in the cassette shell, then fit the ports. Stuart used GPIO sound and an RCA connector for composite video out and added a shutdown button on the front of the tape, and then hooked it up to a shutdown script.

Kimberlina Droid

Selin took inspiration from Star Wars when designing a Raspberry Pi Pico-powered battle bot. She talks strategy with **Rosie Hattersley**

MAKER

Selin Ornek

Selin lives in Istanbul and fits building robots and giving TED talks about STEM around her schoolwork. Her mission is to use AI for good.

selinoid.com

▼ Selin and Kimberlina take to the battleground and prepare to vanquish less speedy foes

I n *Star Wars: Attack Of The Clones*, **a convoy of tanks with enormous wheels that bear down inexorably on their enemies was a fearsome sight**. The concept seemed ideal to 15-year-old robot builder Selin Ornek when she needed ideas for a new battlebot. Having already made quite a name for herself in the world of coding, Selin was likely to have plenty of competitors keen to take her down, which partly explains why she named her battle bot Kimberlina in tribute to a comedy character from TV show *Full House* that she watched in lockdown. Kimberlina may have a quirky name, but her strike is deadly! Once you know something about Kimberlina's maker, this won't be a surprise.

Selin is an accomplished coder and Raspberry Pi robot builder. She began coding aged eight – first Scratch, but later Python, Java, and C++ – and designed her first robot at ten.

Her interest in coding began after Selina interviewed a family friend, who is a mechanical engineer, about building a robot dog as a means of bringing her much-loved dog, Korsan, back

to life. His advice to learn coding and robotics, plus Scratch-based lessons that helped her class learn English, set Selin on her robot-building journey. She has now built six robots, including two that act as guide dog robots for blind people. The original version of this, the Arduino-based IC4U, won Selin her first prize at Coolest Projects International in 2018. "I won first place in the Hardware Category at Coolest Projects, and one of my prizes was a Raspberry Pi 3B+." The other was a Google AIY kit – presented to her by Raspberry Pi's very own Eben Upton. "After this, I started to use Raspberry Pi in all my projects."

With a firm focus on using technology for good and demonstrating that an interest in computing is gender-free, Selin won the Aspiring Teen category in the 2021 Women In Tech Global Awards (**magpi.cc/witawards**).

Firm foundations

Kimberlina began life as a cardboard prototype, which helped Selin decide on the parts she was going to use. The robot seemed a great opportunity to try out Raspberry Pi Pico for the first time, since she needed a microcontroller that could run motors, servos, and receive Bluetooth signals. Selin used the wheel design from her previous robot and designed other parts in Autodesk Tinkercad. "Because the wheel has a very small contact point with the ground, I had to find a way to balance the robot when it accelerates forward and backwards, so I placed steel ball bearings with enough space to move within the robot," she says. The contest in which Kimberlina was to compete threw up challenges too, stipulating maximum weight and speed limitations. "The weight was a challenge after the decision to use ball bearings, but choosing the right motors was the key to [having] a fast-moving and turning robot."

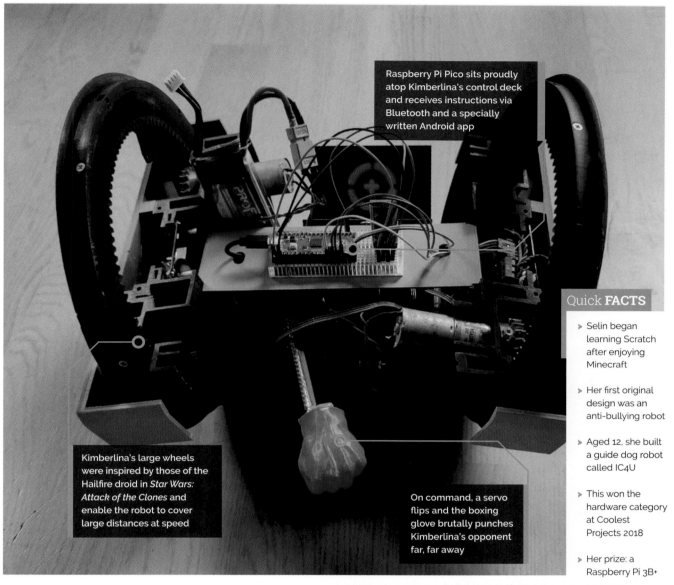

Raspberry Pi Pico sits proudly atop Kimberlina's control deck and receives instructions via Bluetooth and a specially written Android app

Kimberlina's large wheels were inspired by those of the Hailfire droid in *Star Wars: Attack of the Clones* and enable the robot to cover large distances at speed

On command, a servo flips and the boxing glove brutally punches Kimberlina's opponent far, far away

Quick **FACTS**

> Selin began learning Scratch after enjoying Minecraft

> Her first original design was an anti-bullying robot

> Aged 12, she built a guide dog robot called IC4U

> This won the hardware category at Coolest Projects 2018

> Her prize: a Raspberry Pi 3B+

> ❝ Selin's competition strategy was to move quickly and remain agile avoiding contact ❞

Decisions made, Selin 3D-printed and assembled Kimberlina's body, using the time available while the body parts were printing to design a mobile app in MIT App Inventor to control the droid over Bluetooth. She used MicroPython to code Raspberry Pi Pico and the hardware connected to it. Once the coding was complete and the robot was assembled, it was time to test it. "I was lucky," says Selin of how her meticulous planning

Project Showcase

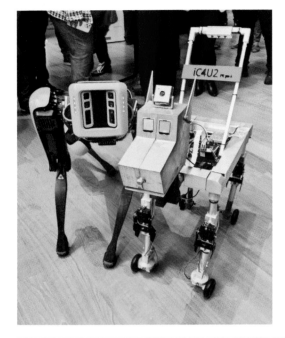

IC4U2 is a Raspberry Pi robot guide dog, one of six robots Selin has designed to date

Having built five other robots, Selin knew exactly what circuitry to use for Kimberlina

A powerful punch and speedy wheels make Kimberlina a fearsome robot opponent

Kimberlina being weighed prior to the competition

Build a bruising battlebot

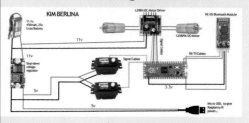

01 Sketch out your design and make a cardboard prototype to help you work out the overall dimensions, how everything will be powered, and the connections needed.

02 Use CAD software to design the robot's body, leaving space for Raspberry Pi Pico and any batteries or cables, then 3D-print or otherwise build the casing.

03 Use the web component of MIT App Inventor to create a means of controlling your battlebot. Set Raspberry Pi's URL as the destination and add **:1880/mit**. An alternative is to use Node-RED and control Raspberry Pi directly.

❝ Kimberlina is controlled via Bluetooth from a mobile application ❞

▲ Selin won first place in the Hardware Category at Coolest Projects in 2018

and experience played out. "Usually my robots do not work immediately. I come across a problem and sometimes even burn a part, but fortunately, Kimberlina worked on the first try."

Speed and control

Kimberlina is controlled via Bluetooth from a mobile application Selin built using MIT App Inventor. A servo at the front controls the flipper to flip an opponent, while the servo at the back is attached to a lever that pushes the opponent away. "Kimberlina has great balance thanks to steel ball bearings that help centre it. This means that when it gets a hit or accelerates fast, it doesn't roll over," explains Selin. "My competition strategy was to move quickly and remain agile, avoiding contact, and make the other robots fall from the platform without touching them," she adds. The mighty wheels of Hailfire, coupled with the portable power of Raspberry Pi Pico, proved an unbeatable combination. 🄼

M4All

Making microscopes accessible and affordable by open-sourcing it. **Rob Zwetsloot** takes a peek

MAKER

Gemma Cairns

A NanoBioPhotonics PhD student at the University of Strathclyde, developing imaging technologies.

magpi.cc/m4all

It's not too hard to understand that Raspberry Pi cameras can be used for microscope builds – you just need the right lenses to focus the image on a small enough object. With a bit of custom code and software paired with it, you can then use it to view a tiny, hidden world.

"We have developed M4All, a fully open-source, 3D-printable, and modular microscopy system, which enables wider accessibility to advanced imaging techniques at a much lower price point than commercial microscopes," Gemma Cairns, one of the students working on this project, tells us. "We integrate Raspberry Pi components into our M4All systems for microscope control, image capture, and data analysis."

The team, the NanoBioPhotonics group at the University of Strathclyde, have been using their imaging expertise to "investigate the mechanisms

▼ Another diatom shot, with another amazing pattern

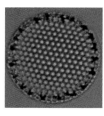

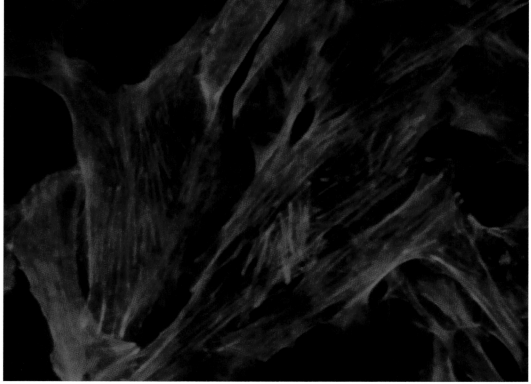

▲ This is actin, a protein that forms the cytoskeleton of cells, illuminated with fluorescence

that macrophages (a type of immune cell in the body) employ to remove respiratory pathogens", pathogens that, among other things, can lead to pneumonia or meningitis.

"From my experience using research microscope facilities to image macrophage samples, I realised there can be limitations with accessing commercial microscope systems, not only with costs but also time limits and flexibility," Gemma explains. "A high-end research microscope can cost over £250,000 and so access to it can be limited and charged per-hour! In the past few years there has been a growing community of research groups working on open-source microscopy hardware and software solutions. Seeing their work inspired us to

> # ❝ We wanted to create a system that can be used by a wide level of expertise ❞

create M4All, which directly addresses our imaging requirements for macrophage research, but also contributes to the community where the system can be adapted for many other applications."

Raspberry Pi microscope

The team opted for Raspberry Pi because they wanted something low-cost and small, and Raspberry Pi's Camera Modules and HQ Cameras, along with its GPIO pins for controlling motors

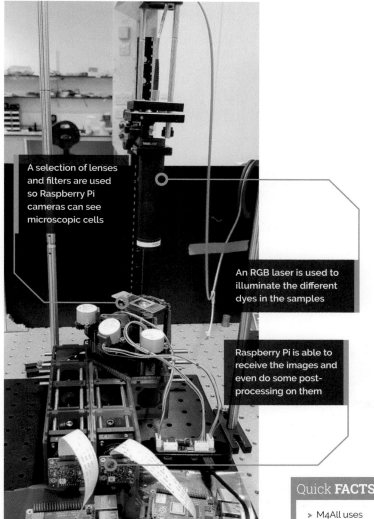

A selection of lenses and filters are used so Raspberry Pi cameras can see microscopic cells

An RGB laser is used to illuminate the different dyes in the samples

Raspberry Pi is able to receive the images and even do some post-processing on them

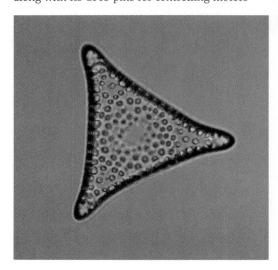

▲ This is a diatom, a single cell of algae which has silica cell walls which create an amazing pattern

and LEDs, made it perfect as the heart of M4All. The next step was just making it all fit together.

"Developing the designs for the 3D-printed parts has been a very iterative process to achieve their final form," Gemma says. "We wanted to create a system that can be used by a wide level of expertise, but also has the stability and capability for quantitative biological imaging (that is, images where the brightness of each pixel can be trusted and compared with nearby pixels or other regions of the image to perform numerical analysis). The modular cubes are printed as monolithic parts to form the optical path, and the inserts are designed to minimise the number of degrees of freedom for alignment. The only optical alignment that needs to be carried out is setting up and focusing the lenses for imaging."

Quick **FACTS**

> ➤ M4All uses fluorescence microscopy that involves fluorescent dyes...

> ➤ ...which are illuminated by different light and then filtered

> ➤ The next stage is testing live cell imaging in a cell culture incubator

> ➤ OpenFlexure was created by the University of Bath

> ➤ The ultimate aim is to see real-time interactions with bacteria and other cells

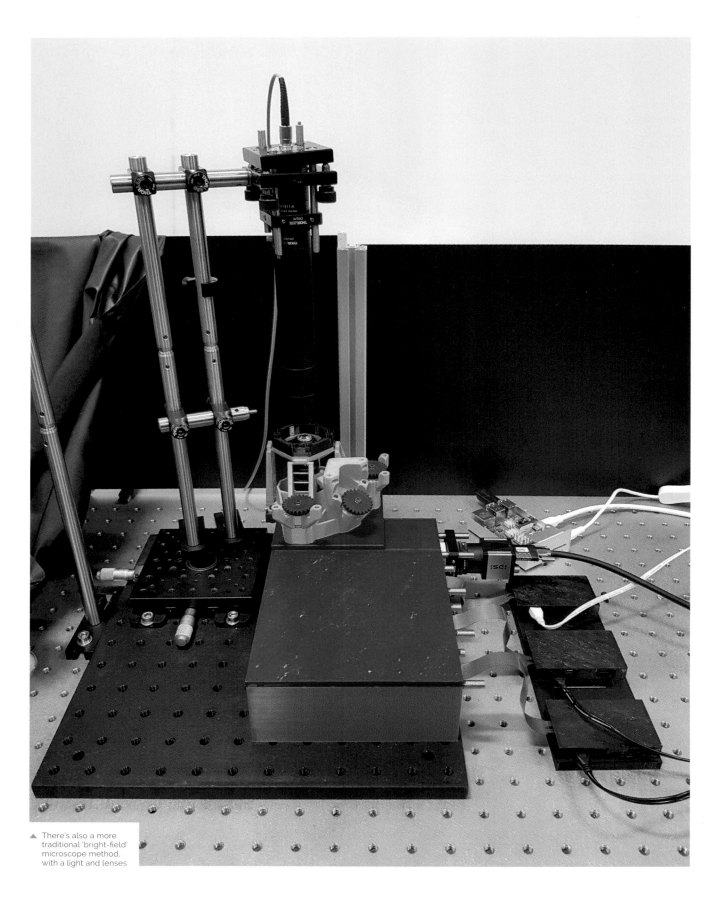

▲ There's also a more traditional 'bright-field' microscope method, with a light and lenses

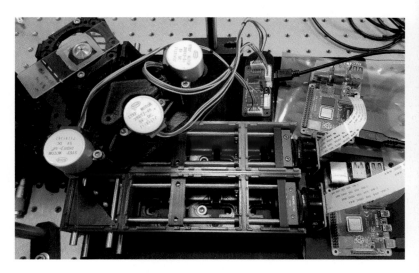

As an open-source project, it also integrates other open-source bits into it, like OpenFlexure. OpenFlexure allows the image to be separated in a specific way and focused on the camera installed in Raspberry Pi.

▲ To get the right focus and filtering, there are a lot of motor-controlled lenses to manipulate

Small goals

"We have been able to image various structures inside of cells using fluorescence microscopy, such as mitochondria, bacteria, and actin, with good resolution and with a signal-to-noise ratio which allows us to obtain quantitative information on what's happening in the cell," Gemma reveals. "This will allow us to study interactions with bacteria inside of macrophages at a much lower cost than commercial systems. Then, if interesting data arises, more specialised microscopy systems can be used for further analysis based on what the M4All microscopes have found. We have also imaged diatoms in bright-field which are single cells of algae. Their cell walls are made of silica, which forms very intricate and cool patterns."

You can head to **magpi.cc/m4all** for instructions on how to build your own system, and start doing your own microbiology.

"Having a flexible platform also means that we can test out new designs and applications in collaboration with a huge range of researchers – from biochemists or marine environmental scientists!" M

See the sights

01 You can choose between two different methods of microscopy – bright-field or fluorescence. You'll need to prepare your setup slightly differently for each style, although both use a similar series of stepper motors and lenses.

02 For fluorescent microscopy, you'll need to add fluorescent dye to the sample. This will react to different wavelengths of light.

03 You can then observe and analyse your results via Raspberry Pi as it captures and processes the image.

Open Weed Locator

Distinguishing between valuable crops and unwanted upstarts requires precise plant knowledge and some Raspberry Pi processing, learns **Rosie Hattersley**

MAKERS

Dr William Salter and Guy Coleman

University of Sydney academics Guy and William developed OWL using Raspberry Pi 4 to make weed management significantly cheaper

■ **magpi.cc/owlgit**

Spring has arrived in the northern hemisphere, along with the weeds that pop up ever more fervently each year. Picking off unwanted plants but leaving others behind requires knowledge and precision – exactly what machine learning is adept at. OWL (Open Weed Locator), developed at the University of Sydney, uses Raspberry Pi 4 to make managing agricultural sites with robots more efficient. It is "a green-on-brown weed detector that uses entirely off-the-shelf componentry, very simple green-detection algorithms and entirely 3D printable parts," explain its makers. The Raspberry Pi 4-based OWL detection system can be mounted on a ruggedised vehicle or tractor and costs a mere $400.

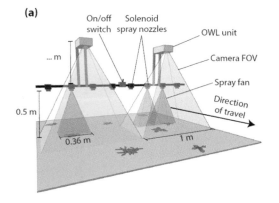

▶ Open Weed Locator sensors ignore anything green and apply fungicide or insecticide to other parts of the crop

Precise planting

Guy Coleman has extensive experience as an agricultural scientist, and began using Raspberry Pi five years ago as a means of exploring how computer vision might be used in such settings. Weed recognition and precision control using deep learning is the focus of his PhD at the University of Sydney. Before this, Guy was more comfortable doing precision weed-control fieldwork on large-scale paddocks in Australia than developing projects using Python such as the OpenWeedLocator.

He works alongside Dr William Salter, whose background is in plant physiology and open-source technology for plant phenotyping, and who had already built several light sensors and an instrument for the high-throughput measurement of photosynthesis.

"Managing weeds in crops so they don't reduce yields is a big challenge in agriculture, and current methods rely on herbicide applications to whole fields," explains Guy. "Being able to assess where weeds are with cameras means the herbicide is only applied to individual weeds, meaning big savings to the farmer and reduced chemical inputs to the environment."

However, weeds vary hugely in colour, size, and shape and the team needed to find a way of recognising them in all sorts of environmental conditions. Since the weed detector also had to work at a reasonable speed, any algorithm used would have to operate with high frame rates, Guy explains. They chose to base OWL around an 8GB Raspberry Pi 4 because of its combination of low cost, high power and small form factor. "Being easily connected to a whole variety of inputs

Raspberry Pi HQ Camera and Raspberry Pi 4 8GB plus a relay board detect weeds and send details back to the robot to act upon

Because OWL detects anything green, it can be used to apply fungicide or herbicide only where it sees a plant, saving on fertiliser costs

OWL was designed to use off-the-shelf components, and runs off a 12v battery

▼ The finished Open Weed Locator build

and outputs has been absolutely essential to this project," Guy comments.

A green detection algorithm running on Raspberry Pi identifies any green weeds that appear in the video feed and then activates a GPIO pin that connects to a relay board. A solenoid can then be switched on to deliver herbicide to the detected weeds.

Smarter applications

Guy and William wrote the code for OWL in Numpy and OpenCV. Keeping OWL open source means it can be easily updated with improved weed detection capabilities. Their biggest challenge was developing an algorithm that performed at an acceptable level in a range of conditions but testing convinced them to settle on a combination of the Excess Green + HSV thresholding systems.

OWL is very much a community project, with the full hardware details posted on GitHub (**magpi.cc/owlgit**) including a 3D-printable enclosure. This has already led to versions of OWL being assembled and used on four different continents, with some tweaking of the enclosure design for easier printing and assembly.

▲ Guy and William with
two OWLs mounted
on the University
of Sydney's Digital
Farmhand robot
at the University's
Digifarm in Narrabri,
seven hours north
of Sydney

▶ OWL can be mounted
on a tractor or the
back of a truck or car

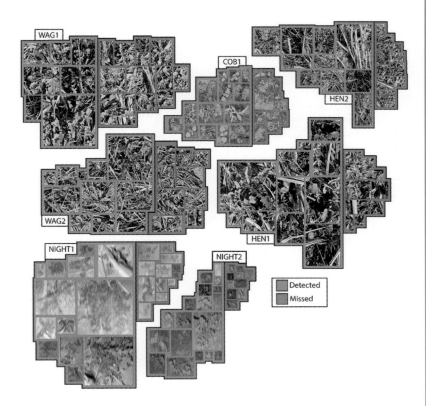

WAG1

COB1

HEN2

WAG2

HEN1

NIGHT1

NIGHT2

- Detected
- Missed

The most critical pieces are the Raspberry Pi 4 with 8GB, Raspberry Pi HQ Camera, a relay control board and 12V to 5V 5A DC to DC converter, says Guy. Aside from printing parts, building OWL takes a couple of hours and running costs are minimal – only 12V of input power required for it to run.

▲ Guy and William used a dual weed detection approach as straw-coloured weeds were often missed by the ExHSV algorithm

" OWL is very much a community project "

Guy and William plan to add in-crop weed detection and GPS and say quite a few farmers have spoken to them about different uses they see for OWL to improve the efficiency of food and fibre production globally. The ability to find anything green means OWL can also be used to only apply fungicide or insecticide to the crop or to defoliate green cotton plants. "OWL is a living project. Now that it has 'flown the nest', so to speak, we're excited to see where the community takes it." M

Make your own OWL

OWL is open source, with full hardware assembly instructions at **magpi.cc/owlassembly**. You can download and 3D print the enclosure too.

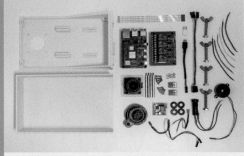

01 Install the owl.img software on a fresh installation of Raspberry Pi OS. Name the virtual environment owl. Assemble the hardware shown in the image.

02 Install Raspberry Pi HQ Camera and enable it with raspi-config. Download the entire OpenWeedLocator repository into Raspberry Pi's Home directory

03 Install the camera as shown, make the Python file greenonbrown.py executable and use bash to make it run at startup. Set up the camera to view the live feed found in greenonbrown.py and reboot.

NOUS: uNdersea visiOn sUrveillance System

Monitoring submerged wrecks using AI and Raspberry Pi reveals the secrets of the deep, writes **Rosie Hattersley**

MAKER

Dr George Papalambrou

George is an assistant professor at the University of Athens' School of Naval Architecture and Marine Engineering, where he is involved in research related to AI and machine learning

▌ nous.com.gr

Watching shoals of fish flit idly by is one of the most magical and calming experiences. Just such an opportunity is a wonderful by-product of Greece's NOUS project, the uNdersea visiOn sUrveillance System. Based on Raspberry Pi 3 and 4, its cameras help researchers from the National Technical University of Athens monitor the submerged shipwreck of the a merchant ship near Peristera, one of the largest known ships from classical antiquity.

Dr George Papalambrou and his colleagues Vasilis Mentogiannis and Kostas Katsioulis, from the NTAU's School of Naval Architecture and Marine Engineering, knew plenty about Raspberry Pi before selecting it for their underwater archaeology surveillance project. "It was our first choice from day one," George says. For a start, he had used Raspberry Pi alongside Apple HomeKit for home automation, and while at university it was used in CAN-bus networks. George was

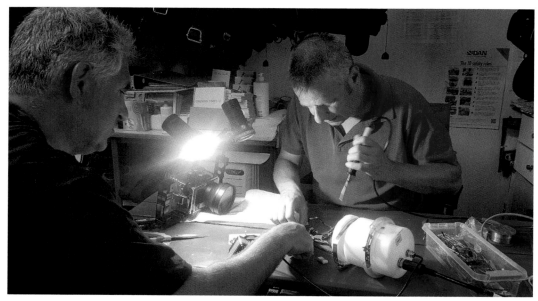

▶ The NOUS team working on the first submarine surveillance camera prototype

Images triggered by AI algorithms and captured by NOUS' camera are sent wirelessly back to the university's servers for analysis

NOUS operates at depths of 35 to 70 metres and is connected by a cable and harness to the research vessel

Raspberry Pi 3 and 4 and an AI camera inside a ruggedised enclosure capture images of the wreck of the merchant ship and its cargo of approximately 4,000 amphorae

Quick FACTS

> NOUS is roughly equivalent to the classical Greek word …

> …which translates to intellect, thinking and reasoning

> The Peristrera was a 25m-long merchant ship

> It sank in the 5thC BCE, on its way to Skopelos

> It is now the world's first underwater ancient archaeological museum

also interested to read about Raspberry Pi Compute Modules being used by the University of Surrey for their Cube-Sats (**magpi.cc/cubesatproject**), confirming the hardware's suitability in challenging environments and ability to communicate while being self-powered.

❝ The sea is an unforgiving environment, and very hard on equipment ❞

NOUS, as the marine surveillance project came to be known, would also need to be able to communicate remotely.

Sea-worthy specifications
George says Greece had wanted to monitor its marine archaeological sites for many years, but one of the main obstacles was how to guard and protect them.

It needed a system that was self-powered (since most wrecks are located a long way from a power supply), that could connect to the internet in order to communicate and be remotely controlled, have sensors to monitor the area of interest

continuously, and be able to send alerts in cases of violations, alterations of the site or other events. As well as monitoring protected marine areas, scoping exercises suggested it would also be feasible to include real-time scientific observations throughout the day, and to monitor changes to the climate and biodiversity in the area over long periods of time.

The sea is an unforgiving environment to operate in, and is very hard on equipment, says George, so it was critical they chose gear that could withstand both high pressure and low temperatures. NOUS needs to run continuously round-the-clock at submerged depths of 35 to 70 metres. George explains that the project also needs total software control at the operating system level, as well as at the application level: "We control our devices remotely over the web and SSH, so there is no space for failures or malfunction." Raspberry Pi was always the team's first choice, not least because of its affordability and the invaluable community forums.

Having bought Raspberry Pi 3 and 4, plus some basic off the shelf electronics, the NOUS team soldered on cables and parts in order to save space in the rugged enclosures that also needed to accommodate AI cameras and

▲ NOUS's dashboard shows live views from each of the undersea cameras as well as logging temperature patterns and local weather conditions

▲ A diver has been detected entering the protected marine area, triggering an alert via Raspberry Pi

Underwater archaeology

01 NOUS primarily uses an 8GB Raspberry Pi 4 at the heart of the AI camera-based marine surveillance setup, along with temperature, moisture and movement sensors, and a Raspberry Pi-compatible camera

02 Sturdy communications, Raspberry Pi running in headless mode, self-contained battery packs and durable waterproof enclosures are also critical elements.

03 The NOUS cameras are positioned near the seabed around the wrecked merchant ship, and are able to capture images at 30fps. Onshore solar panels top up the power supply.

networking hardware that could be attached via a harness and operate underwater.

▲ Camera one of five surrounding the Peristrera, showing Raspberry Pi's camera in use

Academic expertise

The software and specialist HATs were developed by George and his University of Athens colleagues to save money and reduce development time. Raspberry Pi boards were set up headless, with X11 forwarding used to optimise remote control along the lengthy undersea cables connecting each module to the base station. The onboard battery packs are supplemented by onshore solar panels sited near where divers set off to view the wreck in the Aegean, some distance short of its intended destination, the island of Skopelos.

Operating full-time since 2020, the surveillance system is still running successfully today. "Raspberry Pi has been a success since the very beginning, providing stability on both the software and hardware sides", says George.

Live footage from the wrecked ship can be seen at (**magpi.cc/peristera**). M

Szerafin MM5D
Mushroom Farm

Hobbyist farmers Judit and Zsolt used engineering and tech know-how to expand their mushroom farming business. **Rosie Hattersley** hears how

MAKER

Zsolt Pozsár

Judit and Zsolt Pozsár's remotely managed mushroom farm is an excellent first Raspberry Pi project for the engineer and his wife.

magpi.cc/szerafin

With two decades of engineering experience gained during military service, Hungarian Zsolt Pozsár resumed his vocational qualifications and began a career maintaining the bottling lines for a multinational company. Meanwhile, his wife Judit's small farming concern seemed as though it could benefit from Zsolt's knowledge of how technology could improve their crops. The couple have been growing mushrooms for the past five years. Zsolt's MM5D Raspberry Pi-based monitoring system – a four-channel programmable control and remote monitoring system – has kept an eye on the mushrooms' development for the past three years, first in a cellar underneath the house, and more recently in dedicated mushroom fruiting chambers.

▶ The mushroom tents have recently been joined by similarly controlled outdoor vegetable crops

"The aim is to ensure the automatic operation of mushroom growing sites, so the characteristics of the growing environment can be monitored and modified remotely," says Zsolt. "Devices must operate continuously and reliably in a humid environment without user intervention." Remote management without smart controller devices such as relays and timers is impossible. "These operations should be user-friendly, with no programming knowledge required."

Zsolt wanted a small, low-power, Linux-based single-board computer. He chose Raspberry Pi for his mushroom farm on the recommendation of a technical vocational high school teacher who'd used them and said he wouldn't be disappointed. Zsolt now uses Raspberry Pi 3B+ in all his projects, all of which he self-designs, making use of standard items such as display, relay boards, and third-party libraries. The hardware cost no more than 40,000 Hungarian forints (approximately $110, or £90). "The hardest part of building a device is making the box and making the front panel. It required a closed box. I bought this, but it was hard to find a company that deals with Plexiglas cutting and printing," Zsolt explains.

Mushroom for growth

The mushroom monitoring system consists of a TTL-level input in which temperature and humidity are measured. The remotely accessible

Logging in via SSH allows the couple to check for any issues with overwatering or humidity, as well as optimising light levels

Zsolt's MM5D controller is based around Raspberry Pi 3B+, and runs a bespoke Python script

The MM5D remotely controlled mushroom growing environment is successful enough to operate as a commercial farm

Quick **FACTS**

> Zsolt enjoys P languages: Pascal, Perl, and Python

> And has a weakness for rescuing abandoned cats

> His army years were spent looking after air defence radars

> He still can't resist repairing old electron tube equipment

> But he's currently tinkering with irrigation controls

Warning!
Mains Electricity

Do not connect 230 V contacts to the relays on the printed circuit board, as their high coil voltage and current will also cause the contact to spark and interfere with the display's serial data traffic. Use low voltage and current fast relays and move them away from the control unit.

magpi.cc/ electricalsafety

▲ Mushrooms benefit from their optimal growing conditions

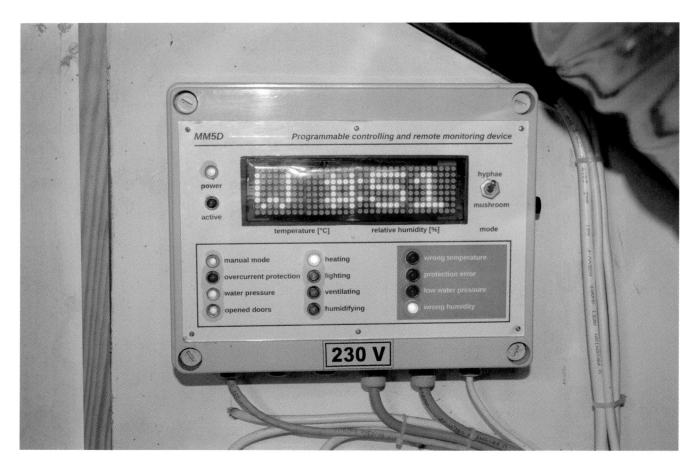

▲ Zsolt's controller allows the user to see at a glance the status of the crop and the current environment settings

system shows the status of the lights, fans, humidifier, and watering system – including the pressure, and whether or not the tent door is open. "Measurement, timing, and electrical equipment control is done by a Python-language program that

> ❝ Growing mushrooms this way has proved successful enough for Judit and Zsolt to operate a local delivery service ❞

runs as a service in the background," Zsolt tells us. The program requires an internet connection and access to data from **OpenWeatherMap.org**.

The MM5D setup (one of several he's developed and implemented at the family farm), uses Raspberry Pi 3B+ and has been in continuous use since 2019. "The devices were built one after the other, so I was able to use the experience gained in building for the next device." This iterative process is reflected in Zsolt's meticulous GitHub

where he makes the installer software for his MM5D plant-monitoring devices available online (**magpi.cc/mm5dgit**). He also maintains his own Debian repository for Raspberry Pi OS, Debian, and Ubuntu Linux (**magpi.cc/pozsi**).

Grow more greens
Growing mushrooms this way has proved successful enough for Judit and Zsolt to operate a local delivery service for customers in the vicinity

of their farm in Tiszaföldvár, close to Budapest. Their mushrooms, as well as oyster mushroom compost useful for growing other crops, are sold in environmentally friendly packaging.

They've also expanded with additional mushroom fruiting chambers, as well as diversifying into other crops. The addition of these vegetable plots has also given Zsolt the excuse to come up with another monitoring project adding automated irrigation to the whole site, as well as the mushroom tents. "The water and electrical system, electrical cabinet, and pump shaft will be ready by the summer. Tomatoes, eggplants, and pumpkins will grow and the environment will be beautiful!"

▼ Judit and Zsolt's son shows off the latest mushroom crop grown in their cellar, produced with the help of the original 2018 prototype controller

Grow your own

01 The MM5D controller unit houses Raspberry Pi and sensors and controllers to adjust the lights, temperature, and humidity sensors. It's linked to **OpenWeatherMap.org**.

02 Powered by Raspberry Pi 3B+, the MM5D has four inputs, four controllers, and four status relay contact outputs. Parameters are customisable. Details of the Python-based hardware controller are at: **magpi.cc/mm5d**.

03 Environmental settings can be manually overridden, but the system is largely designed to operate automatically.

BirdNET-Pi
bird identification
listening station

Curiosity about local wildlife led to the creation of BirdNET-Pi, which identifies birds by their song. **Rosie Hattersley** quizzes its founder

MAKER

Patrick McGuire

Patrick McGuire is a 'nature-loving tinker', citizen scientist and accidental tech enthusiast.

magpi.cc/birdnetpigit

Patrick McGuire describes his BirdNET-Pi acoustic monitoring project – making use of Raspberry Pi to achieve it – as a contribution to making citizen science more accessible and affordable. The approach also chimes with his own route to discovering the open-source community: he first became interested in tech when tasked with fixing his partner's seemingly 'dead' MacBook which was long out of warranty. Having successfully revived the laptop by taking a punt on installing Linux (with the wealth of useful information about it online, he reasoned it couldn't be too problematic to set up and use; he laughs at this assumption, but the 'dead' MacBook is still in use), Patrick began exploring more computing projects after being given a Raspberry Pi as a present. Since then, he's been using Linux and Raspberry Pi to "solve my problems, satisfy my curiosities, connect me with others, and contribute to citizen science."

Credit where it's due

BirdNET-Pi is based on the Lite version of the eponymous BirdNET research platform created by Stefan Kahl and others at the K Lisa Yang Center for Conservation Bioacoustics at the Cornell Lab of Ornithology and the Chair of Media Informatics at Chemnitz University of Technology. Their research is mainly focused on the detection and classification of avian sounds using machine learning, and works on most computing platforms, as well as having iOS and Android apps. The BirdNET database recognises around 3000 species of birds, around 1000 of which are European species.

Patrick's version, BirdNET-Pi, expands the user base for the acoustic monitoring platform, using Raspberry Pi OS for on-device analysis and processing. He thinks the way "BirdNET-Pi automatically creates clips of detected bird songs with spectrograms to visualise the audio

Raspberry Pi running BirdNET-Lite analyses bird sounds, checks them locally against its database, and logs the species' presence

BirdNET-Pi works with almost any USB microphone and sound card, automatically recording birds singing

An Ethernet cable running from the acoustic listening station means it can be positioned in a wildlife-rich location, but bird calls can be reliably logged and details sent to BirdNET

Quick FACTS

▶ BirdNET-Pi began as a means to impress Patrick's partner's mother

▶ The logo is a northern cardinal, the official bird of Virginia

▶ BirdNET-Pi is built on the TFLite version of BirdNET

▶ BirdNET-Pi recently caught the imagination of UCL...

▶ ...which ran a sold-out workshop on it in June

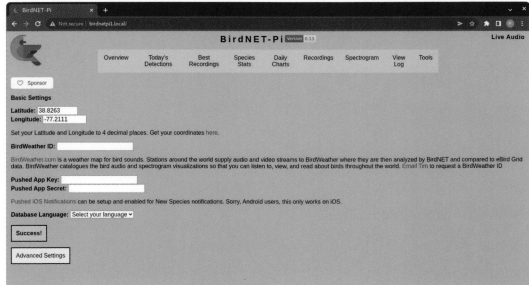

You can specify the listening station's exact location among other settings

The BirdWeather map shows species recently detected by BirdNET-Pi and other acoustic monitors

Identify bird sounds

01 A waterproof housing, Raspberry Pi, microphone, fan, and cables make up the basics of an in-the-field BirdNET-Pi station. Photos from Bill Powers' setup (**magpi.cc/birdnetbuild**).

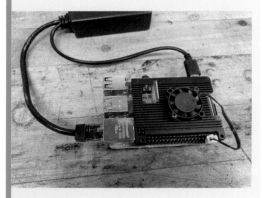

02 The Power over Ethernet card and PoE adapter, plus a power supply with fan, ensure BirdNET-Pi can be used to remotely record and identify bird species.

03 The dashboard view for BirdNET-Pi shows a bird call has been recorded, analysed, and checked against the songbird database, revealing that a pileated woodpecker has been detected.

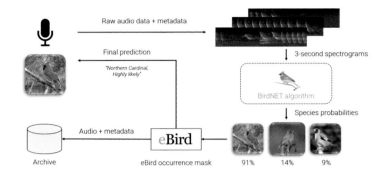

Raw audio data + metadata

Final prediction

"Northern Cardinal, Highly likely"

3-second spectrograms

BirdNET algorithm

Species probabilities

Archive

Audio + metadata

eBird

eBird occurrence mask

91%   14%   9%

▲ BirdNET uses sophisticated AI algorithms to predict which bird is singing

and each bird's unique vocal signature" is a game-changing feature.

The original BirdNET was trained in the Sapsucker Woods close to the Cornell Conservation Bioacoustics Labs. BirdNET-Pi was developed in Virginia, where Patrick lives, and is one of dozens of listening stations that have been set up and left in situ with a wireless connection allowing for remote bird monitoring. "Since all of the analysis is performed on the device itself, researchers can deploy their BirdNET-Pis in places of avian interest where human monitoring for the same duration would be impossible – they can come back later to retrieve the device and all of its analysed data."

BirdNET-Pi also uses Icecast (**icecast.org**) for live audio-streaming, so fellow nature watchers can tune in and hear recordings of live birdsong from other BirdNET devices. There are a handful of broadcasting stations in the UK, plus dozens of other sites you can visit virtually to hear bird song recordings and learn about the species.

Based on the BirdNET-Lite bird sound analysis model, it works on a Raspberry Pi 3B+, 4, or

▼ Recording of a successfully identified chimney swift singing

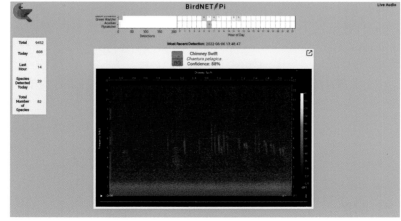

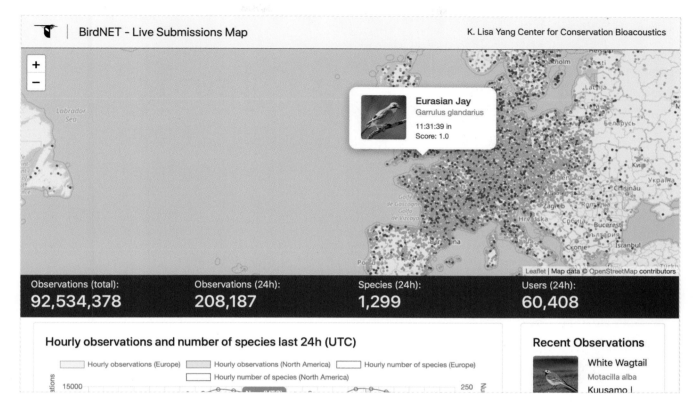

BirdNET - Live Submissions Map — K. Lisa Yang Center for Conservation Bioacoustics

Eurasian Jay
Garrulus glandarius
11:31:39 in
Score: 1.0

Leaflet | Map data © OpenStreetMap contributors

Observations (total):	Observations (24h):	Species (24h):	Users (24h):
92,534,378	208,187	1,299	60,408

Hourly observations and number of species last 24h (UTC)

Hourly observations (Europe) Hourly observations (North America) Hourly number of species (Europe)
Hourly number of species (North America)

15000   250

Recent Observations

White Wagtail
Motacilla alba
Kuusamo L

Zero 2 W, requiring the 64-bit version of Raspberry Pi OS. Raspberry Pi was chosen for its affordability, and Patrick is particularly proud of the fact BirdNET-Pi works with such a range of USB sound cards and microphones. "This allows users to use hardware that fits their needs and their budget, lowering the barrier of entry for participation."

Raspberry Pi proved a great choice. "The original version, BirdNET-system, only ran on x86 machines. It could record up to four hours each day and spent the other 20 hours of the day analysing the data! When the project migrated over to using the BirdNET-Lite model, I was able to port everything to the AARCH64 architecture and BirdNET-Pi was born."

Global reach

Although BirdNET (and BirdNET-Pi) was developed and AI trained in the US, it's now being used by ornithologists and nature lovers to identify species across the globe, which means the database of confirmed bird sightings and sounds has broadened significantly. Patrick lists 18 countries in which wildlife enthusiasts are actively using BirdNET-Pi, from New Zealand to Romania, to South Africa, Canada, and Sweden, as well as the UK, Germany, and USA. Along with the iOS and Android app versions of BirdNET, this Linux-based Raspberry Pi version serves to expand the geographical reach of the bird ID project. 🐦

> **It's now being used by ornithologists and nature lovers to identify species across the globe**

▲ A weatherproof box protects Raspberry Pi out in the field

Pico Railway Clock

Giving new life to a clock from yesteryear, **Nicola King** celebrates
the relaxing and reassuring sound of classically cool timepieces

MAKER

Martin Spendiff & Vanessa Bradley

Martin is a mathematical modeller who left the UK for Switzerland, and a fan of FOSS and tech that serves users, rather than the people who made it. Vanessa is new to coding and a constant source of weird and good ideas.

veeb.ch

Readers may remember a fantastic Teasmade project that we featured back in *The MagPi* issue #114 (magpi.cc/114), made by Swiss-based team Martin Spendiff and Vanessa Bradley. Well, while browsing their YouTube channel, we noticed that – in a new upcycling project – they have taken a vintage railway clock and transformed it with a Pico. Naturally, we wanted to know what made it tick.

Old timer

"We bought a railway station clock from a flea market and were a bit crestfallen when nothing happened when we plugged it in," explains Martin. "The nice man who sold it seemed adamant that it worked, so after a bit of reading, I found out that it was waiting for a signal from a 'Mutteruhr'."

As the duo explain in their YouTube video (**magpi.cc/railwaystationclock**), often when you see such clock in a station setting, there is a delay between the second hand reaching 12 and the minute hand advancing; this is because the clock is waiting for an electrical pulse from the 'Mutteruhr', or mother/master clock. This pulse drives the minute hand forward and then the second hand is free to complete another cycle.

Martin and Vanessa had purchased what was essentially a secondary clock – ineffective without a mother clock. To get it working, they decided

to build a mother clock themselves with a few additional components and some code running on a Raspberry Pi Pico microcontroller.

Inner workings

Attaching a ferrite antenna – to pick up the DCF77 atomic clock long-wave radio signals in Europe – to Pico was a first step, along with incorporating a real-time clock (RTC). "There is a text file that tells the code the time that is showing on the clock," explains Martin. "You enter that manually. When Pico is plugged in, the code checks if the recorded time is the same as the RTC time – if not, it sends a pulse to the clock, updates the recorded time by a minute, and says 'what about now?' It just keeps doing that."

The pair encountered few issues during the build and think it would be a relatively easy make to replicate. "The Python code needed to be tweaked a little, but it was relatively plain sailing," says Martin, who reveals that they have now updated the code to also work in the US using the WWVB signal. A radio signal is not essential, however: "Setting the RTC manually would get it to work. The only difference is that the code would not update the RTC."

The 3.3V output from Pico's GPIO pins is converted to 24V by a step-up module, before being routed to an H-bridge to send the pulse to

▶ Wiring up the circuit. Components include an H-bridge, step-up module, and radio antenna

The old railway clock requires a pulse from a mother clock, in this case, a Raspberry Pi Pico

Pico is connected via a step-up module to an H-bridge board to send 24 V pulses to the clock

Connected to Pico, a ferrite antenna picks up the atomic clock radio signal for ultra-accurate time

the clock, although the voltage will depend on the timepiece used. "Some bigger clocks need a bigger electrical 'kick'," notes Martin.

" I found out that it was waiting for a signal from a 'Mutteruhr' (mother clock) "

The pair's interest in old clocks has led to quite the collection. Indeed, Martin admits they now have "too many", but he has a cunning plan to free up some wall space, as he says some will likely become birthday presents "for people that were foolish enough to look interested as we explained to them how they worked!"

As for upcoming ventures, they are certainly not short of spare horological parts. "We've got a box with 74 clock movements in it," reveals Martin. "We mentioned in the video that people often take the original movements out and replace them with quartz movements – we found one of those people, and convinced him to give us the leftovers." They are not entirely sure what they will do with all their clock components: "ideas welcome!"

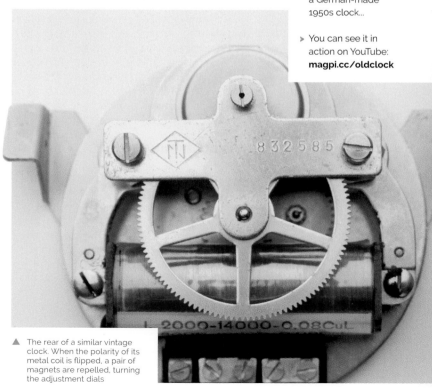

▲ The rear of a similar vintage clock. When the polarity of its metal coil is flipped, a pair of magnets are repelled, turning the adjustment dials

VK-Pocket camera

Long-time Raspberry Pi Zero fan James used one to create a Blade Runner-inspired camcorder, catching **Rosie Hattersley**'s eye

MAKER

James Brown

James lives in Wellington, New Zealand, has a background in games and graphics programming, and creates interactive exhibitions.

@ancientjames
@mastodon.
social

Alert!
CRT

This project uses a CRT television. Be careful of high voltages when working with CRT screens.

magpi.cc/crt

There's "something wonderfully unsettling" about being stared at by your own disembodied eye, comments master of understatement James Brown.

James came up with the idea for his VK-Pocket Camcorder while working on the face-tracking feature for an interactive exhibit, and realised one of the debug tools was showing parts of his face as it detected them. "I had a little viewfinder CRT salvaged from a junk shop camcorder, so I decided to wrap up that experience in a little self-contained gizmo, and style it after the Voight-Kampff machine from *Blade Runner*."

"In the movie, there's a camera on a stalk which is aimed at the subject's eyeball, and a monitor showing that eye isolated and magnified. My concept was to have a high-resolution, wide-angle camera, and use the face tracking code to crop and zoom in to any eye it detected. Anyone approaching the machine to look at it would be stared back at by their own eye."

Animated response

James knew immediately that he wanted to use Raspberry Pi Pico for his VK-Pocket camera project. Moreover, composite video out, which Pico supports, was essential for driving the CRT (cathode ray tube) display he culled from an old video camera. "Raspberry Pi Pico was my first choice for this build. I love these things", he exclaims! "They're a full Linux PC in a microcontroller form factor. I've put them in all sorts of builds, from animatronic heads to robotic insects." [Yes, we want to hear more about these projects, too – Ed].

James is a stickler for details so, as well as accommodating the mini screen, camera, and Pico, it was vital that the case for the homebrew VK machine looked like the original film prop.

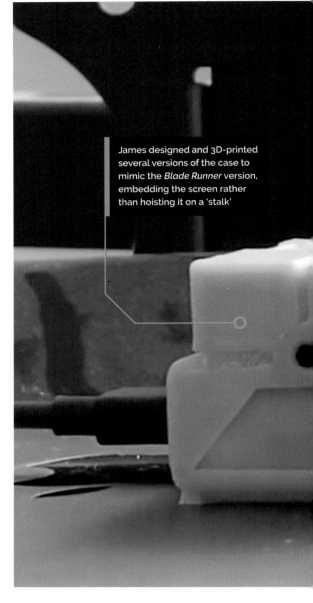

James designed and 3D-printed several versions of the case to mimic the *Blade Runner* version, embedding the screen rather than hoisting it on a 'stalk'

Illustrating this is the "little servo" he added "to push some cosmetic bellows up and down," as a nod to those in the film. There are two versions of the VK machine in *Blade Runner*, he explains; "the device I ended up building is a bit of a mix of both of those, in order to fit everything in."

The servo is controlled using the pigpio library directly from a GPIO pin. Both servo and display draw less than 500 mA, and are powered from the same USB connection so they can be powered from the Pico, with no extra power source needed.

Since it was 3D-printed, James was able to experiment with a few iterations before settling on a design in which everything fits comfortably in place. Even so, he says, the control board for the display ended up at a bit of an odd angle.

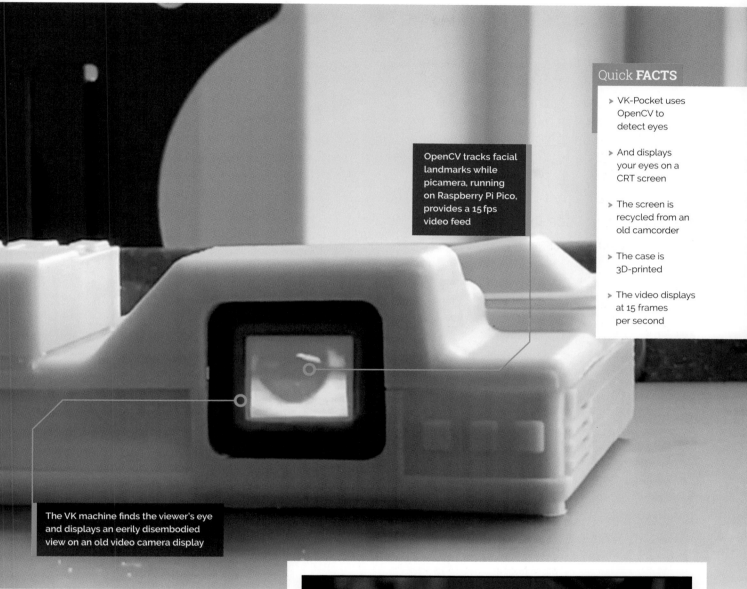

OpenCV tracks facial landmarks while picamera, running on Raspberry Pi Pico, provides a 15 fps video feed

The VK machine finds the viewer's eye and displays an eerily disembodied view on an old video camera display

Putting the camera on a stalk turned out to be tricky, too, "so I put it inside the main case, looking out through a hole."

The eyes have it

James wrote "a quite minimal" amount of Python code (**magpi.cc/pieyepy**) "to keep the high-res live video updated via the GPU while the CPU does the eye tracking." He used OpenCV to detect faces with five facial 'landmarks', from which eye locations are taken. Although the eye-tracker appears to work in real-time, James realised it would be sufficient to have second-by-second updates. "If you wanted to get clever, you could use motion vectors from the compression hardware to improve tracking between detections,

▲ The project was inspired by the tense interview scene in *Blade Runner*

▲ James' Pi Zero animatronic heads are "not quite a robot army", but are very impressive

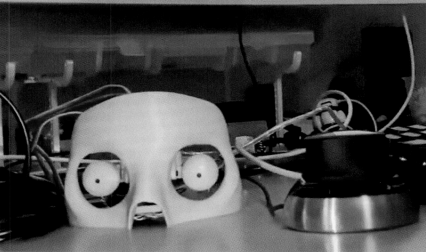

but it seemed good enough just updating every second or so." This reduces the processor overheads and works nicely on a Pi Zero.

The Pico CPU outputs 320 × 240 images at "maybe a couple of frames per second", while the picamera library keeps the screen updated with the live image. "The video hardware can handle 2592 × 1944 at 15 fps, and crop, scale, and display

▲ James' skull has eyes with embedded cameras that swivel and stare unnvervingly at their prey

> ❝ Raspberry Pi Pico was my first choice for this build. I love these things! They're a full Linux PC in a microcontroller form factor ❞

it without touching the CPU, James explains. As a result, the eye region is still reasonably detailed, even though it's only a tiny portion of the camera's view. "If you sit still, it locks on to your eye in less than a second, and stays well centred."

There's no word yet from James on whether his VK-Pocket machine actively analyses its subjects' eyes to check whether or not they may actually be a replicant. ◩

Go eyeball to eyeball

01 To recreate the Voight-Kampff machine, you'll need a Raspberry Pi Pico, HQ camera, a suitable enclosure, and the CRT display from a camcorder. James designed and 3D-printed a case for his.

02 A servo drives mini bellows, like those in the original film. The servo is controlled using the pigpio library directly from a GPIO pin. No extra hardware is needed.

03 Download and install the eye-tracking code from James' GitHub at **magpi.cc/pieyepy**. This will seek out a face, analyse the image, and crop into the eye area, for display on the CRT.

Big Mouth Billy Bass

An attention-seeking interactive fish gets a Pico W update
and becomes an online star, discovers **Rosie Hattersley**

MAKER

Kevin McAleer

Kevin McAleer makes robots, brings them to life with code, and makes videos about them on YouTube.

**magpi.cc/
kevinmcaleer**

I
n the early 2000s, Big Mouth Billy Bass – a kitschy, 3D plastic fish mounted in a picture frame that appears to sing, as well as writhe around – became such an in-demand item that there's reputedly one hanging proudly above the grand piano at the Queen's Balmoral residence. YouTuber, and Raspberry Pi enthusiast, Kevin McAleer relates this apocryphal tale while introducing his latest Raspberry Pi project:

a Pico W–controlled update to the tuneful animatronic sea creature, hosted at **mouthpi.co**.

Kevin first fell in love with computing when he became the proud owner of a ZX Spectrum back in 1982. He went on to study computer science. He had a similarly Damascene encounter when he got his first Raspberry Pi not long after it first launched. "Raspberry Pi has helped me learn and master Linux and inspired me to learn Python,

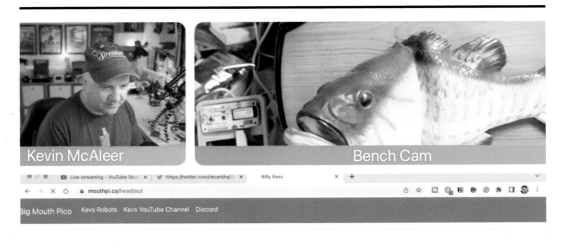

▶ The site being constructed during Kevin's weekly YouTube broadcast

The animatronic fish used to play *Don't Worry, Be Happy*, while flapping his fins

Kev used Raspberry Pi Pico W to make Billy Bass controllable from a web page

Visitors to **mouthpi.co** choose whether Billy should open or close his mouth, move his head or tail, with audio buttons soon to be added

BIG MOUTH BILLY BASS

Quick **FACTS**

➤ Catch Billy Bass in action to see the full effect

➤ Kev contemplated livestreaming footage of Billy writhing around

➤ The current version goes through 9V batteries "very" quickly

➤ So Kev will switch to mains power fish frolics soon

➤ Rob interviewed Kevin for *The MagPi* in issue #118

which is now my go-to language for all projects." Every Sunday, Kevin hosts a YouTube series (head to **magpi.cc/kevinmcaleer**) discussing all

> ❝ Raspberry Pi has helped me learn and master Linux ❞

things Raspberry Pi, and is also an accomplished robot builder.

Big Mouth strikes again

When Pico W launched in June, Kevin was keen to put the wireless-enabled microcontroller through its paces. Several Pico W web-page-control projects appeared online, but Kevin felt they didn't show the new product's full abilities. He'd previously bought a Big Mouth Billy Bass from eBay for around £20, and reasoned pairing it with Pico W might help him "stretch its capabilities beyond common expectations."

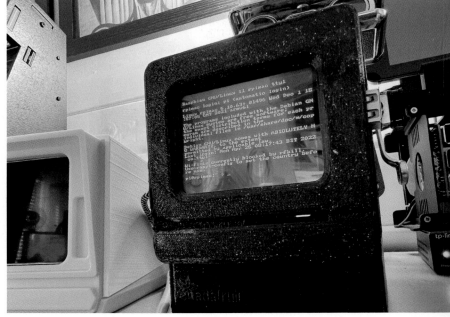

▲ Kevin wanted to show that a Pico W-powered web page could be more than plain text

▲ As a child of the 1980s, Kevin couldn't resist designing a Raspberry Pi-based *Ghostbusters* wireless scanner

He wanted to see how well it would hold up with thousands of web page requests per day, and to see how well the web pages could handle colour, fonts, and style sheets. "Pico W doesn't have an OS and has minimal memory, so being able to host a website and control a robot simultaneously is quite remarkable," says Kevin. The **mouth.pi.co** site is hosted on the Pico W, which is in turn hidden within the fish robot's body.

Kevin contemplated livestreaming footage of Billy writhing around, but the current iteration of the site has buttons that the user can press to initiate preset movements relating to the head, tail, and mouth. A replacement for the audio files containing the original Big Mouth Billy Bass theme tunes – *Don't Worry Be Happy* and *Take Me To The River* – is planned for the next version. Kevin has also promised his YouTube followers a Furby-based Pico W-controlled site.

Hacking the hardware

One of the key aspects of this project was establishing how the existing animatronic fish worked. Online research revealed some details,

▲ The WiFi scanner spreads its arms to show signal strength

usually with a view to controlling the fish with Alexa, whereas Kevin's plan was to control the motors himself. However, a tear-down of Billy Bass's components, in which Kevin stripped out the existing wiring, showed a relatively simple circuit with three motors.

Having learnt these were "cheap 5V DC motors", Kevin was confident he'd be able to drive them with a couple of L298N H-bridge modules. He secured them along with Pico W on a mounting

Big Mouth Billy Bass

This website is hosted on a **Raspberry Pi Pico W**, in the Robotlab, hotglued inside a Big Mouth Billy Bass. You can remotely control Big Mouth Billy by clicking the buttons below.

Catch the YouTube video about this project by clicking here.

[Move Head Out] [Move Head In] [Open Mouth] [Closed Mouth] [Move Tail Out] [Move Tail In]

plate to hold them in place. These would allow him to control powerful motors simply by making a GPIO pin on the Pico high or low (1 or 0). Code shared by Raspberry Pi's Alasdair Allan for making an LED light up came in useful here, as did the

▲ When website visitors click a button to change Billy's pose, the relevant photo is shown

Kevin was confident he'd be able to drive them with a couple of L298N H-bridge modules

realisation that, as well as sharing the 9 V battery between the motors, the battery ground needed to be connected to the Pico W too.

The entire setup cost approximately £20, with a further £20 for the domain name and Cloudflare-hosted website (which offers DDoS protection) covering the next five years. Full MicroPython code and setup instructions are at **magpi.cc/bigmouthwifi**.

Meanwhile, Kevin is already well on his way to his next Pico W project: a *Ghostbusters* PKE WiFi scanner that moves its arms to indicate the strength of the available wireless connection. ▥

In a flap

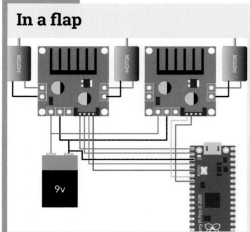

01 Source a Billy Bass or a similar robot. Disassemble it to see what's inside and sketch out the proposed schematic. Here, H-bridges are used to drive the robot's three motors.

02 Kevin used Pico W and MicroPython to set up a 'billy' class, so moving the head and tail and opening and closing the mouth requires simple commands such as 'billy.open_mouth' or 'billy.flap_tail(3)' to flap the tail three times.

03 Prior to reassembling the Billy Bass case, Kevin used a Pimoroni Pico Explorer to test the wiring. The top white motor moved the head, as expected, while the others made the tail flap.

Fireballs Aotearoa

MAKER

Dr James Scott, Jeremy Taylor, Jim Rowe

Dr James Scott, a geologist at the University of Otago, and Jeremy Taylor set up the Fireballs Aotoreoa meteor tracking programme with input from Jim Rowe of the UK Fireball Alliance.

fireballs.nz

A chance to expand meteor monitoring to New Zealand was made far simpler with the aid of Raspberry Pi. **Rosie Hattersley** reports

The excitingly named Fireballs Aotearoa (fireballs.nz) is an ambitious research project that aims to make good on the Global Meteor Network's aim of ensuring no meteor is undetected. The GMN's worldwide meteor and meteorite tracking endeavours already had sites spread across Europe and the US, but few in the Southern Hemisphere. With international news coverage of a fireball over southern England in March 2021 that was successfully tracked by citizen scientists, there has been a significant increase in the number of meteor cameras in the UK and beyond. A Fireballs Aotearoa outreach programme involving New Zealand-based astronomers and meteor researchers generated much excitement among would-be space scientists.

This online presentation given by Jim Rowe (**magpi.cc/findingmeteorites**) explains how a camera attached to a Raspberry Pi locks on to a fireball as it travels across the sky, focusing on the object itself and only processing data relating to its constantly changing location. Raspberry Pi has sufficient processing bandwidth to live-track and report the meteor's location. Since multiple cameras across a region track the same meteor's journey, it's possible to triangulate its final destination and work out with some accuracy where it must have landed, and potentially recover it, for further study, explains Jim.

> " Raspberry Pi and Sony camera lenses have become the de facto hardware "

New horizons

Raspberry Pi and Sony camera lenses have become the de facto hardware, lowering the cost of setting up a meteor camera to less than £200. Expanding coverage in Scotland and New Zealand is important for Jim Rowe of the UK Fireball Alliance. Jim was incredibly excited

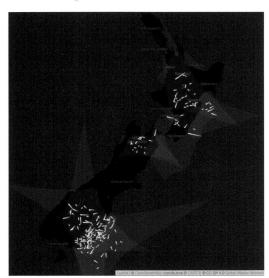

▲ The meteor camera network will monitor the skies over the whole of New Zealand

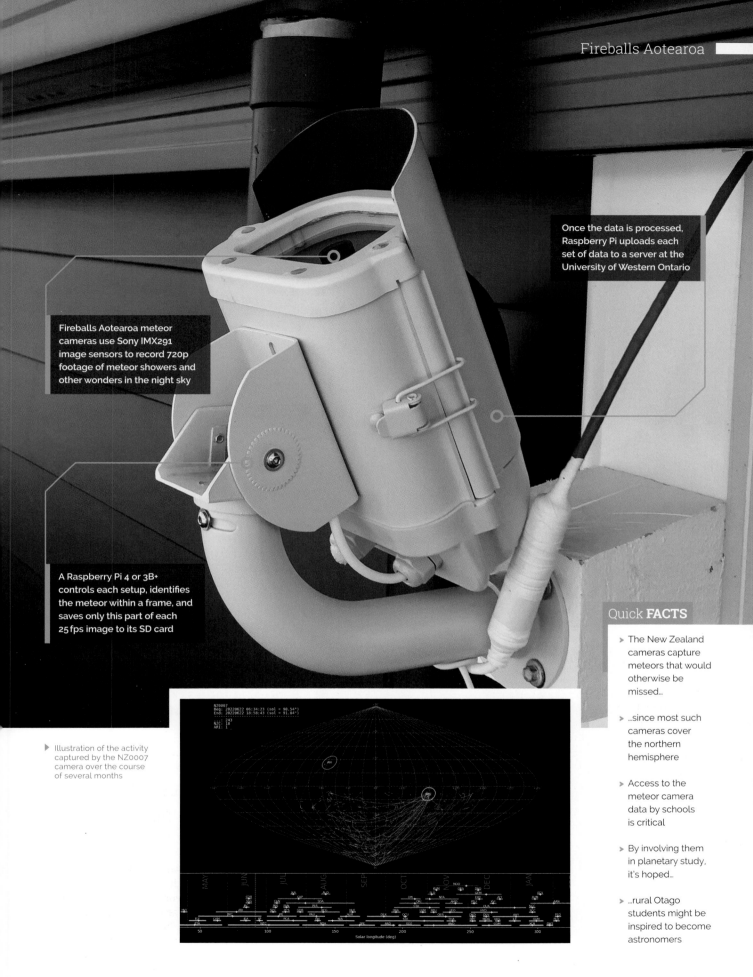

Once the data is processed, Raspberry Pi uploads each set of data to a server at the University of Western Ontario

Fireballs Aotearoa meteor cameras use Sony IMX291 image sensors to record 720p footage of meteor showers and other wonders in the night sky

A Raspberry Pi 4 or 3B+ controls each setup, identifies the meteor within a frame, and saves only this part of each 25 fps image to its SD card

Illustration of the activity captured by the NZ0007 camera over the course of several months

Quick **FACTS**

> The New Zealand cameras capture meteors that would otherwise be missed...

> ...since most such cameras cover the northern hemisphere

> Access to the meteor camera data by schools is critical

> By involving them in planetary study, it's hoped...

> ...rural Otago students might be inspired to become astronomers

◀ The project is already detecting plenty of meteor activity in the southern hemisphere, as this open-source meteor map reveals

RMS NZ0002   03-05-2022

Detected meteors: 000110

▼ UK Fireball Alliance's James Rowe's talk on meteor tracking showed how Raspberry Pi-powered cameras helped triangulate locations

when Dr James Scott of the University of Otago succeeded in getting funding for the first ten meteor camera kits for use in New Zealand, and helped acquire 20 more Raspberry Pi computers for meteor tracking and university use with a grant from the MBIE Curious Minds Participatory Science Platform. A key part of the funding pitch is the direct link with schools and their access to the project data. James says the goal is to record meteors on every clear night – a single camera in Invercargill picked up 114 meteorites crossing the sky one evening in early March – with students able to log in at any time to see what has crossed 'their' night sky and incorporate that in their schoolwork.

"Rather than simply recording pictures of meteors, the idea is to collect science-grade data that can inform researchers, capturing information about meteor orbits, frequency, flux, mass indices, source regions, and so on. This can be used to refine prediction models and help us learn more about parts of the solar system nearest to us. All the high-level data products from the GMN project are publicly released under CC BY 4.0 and updated every six hours so researchers can have access to near-real-time information," explains Jeremy Taylor (aka Tasmanskies) who has been "a driving force" behind the meteor camera builds.

Searching for a sky fall

Only nine meteorites have been discovered in New Zealand, and only the 1908 Mokoia meteorite in Tauranga was seen to fall. With the country having a land-mass larger than the UK, Dr Scott aims "to discover the next meteorite

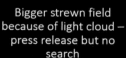

South Wales – 11/12 May 2022

Data from FRIPON and Desert Fireball Network, trajectory by Dr Hadrien Devillepoix, Curtin University WA

Bigger strewn field because of light cloud – press release but no search

Google Earth

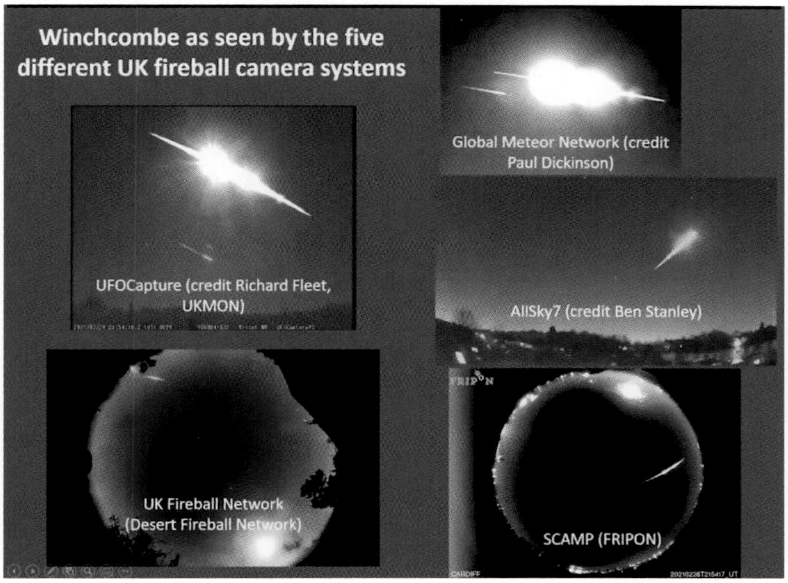
Winchcombe as seen by the five different UK fireball camera systems
UFOCapture (credit Richard Fleet, UKMON)
Global Meteor Network (credit Paul Dickinson)
AllSky7 (credit Ben Stanley)
UK Fireball Network (Desert Fireball Network)
SCAMP (FRIPON)
FRIPON

▲ Multiple camera systems are cross-referenced to aid the recover of falling meteors

that comes into New Zealand through a citizen-led initiative." The meteor cameras are put together by students from the rocketry club at the University of Otago and installed at locations around New Zealand, creating the densest southern hemisphere meteor-tracking network. The network is already capturing plenty of activity. "Raspberry Pi is essential for calculating the meteor trajectory each camera picks up and determining the 'strewn' field where debris should have landed," says James.

> Raspberry Pi is very easy to program and operate. We easily link to the boards at the schools via a remote connection

"Raspberry Pi is awesome; small enough to sit around unobtrusively, but powerful enough to control a night-sky camera and manipulate the data that it collects, such as generating stacked images of the duration of the night," enthuses James. "Raspberry Pi is very easy to program and operate. We easily link to the boards at the schools via a remote connection, which enables us to see the live stream, edit images, and access photos and videos. These are easily recompiled into time-lapses. It's just brilliant."

As a future development, Jeremy Taylor hopes it will be possible to install meteor cameras in Antarctica – with Raspberry Pi inside, of course!

Build a meteor camera

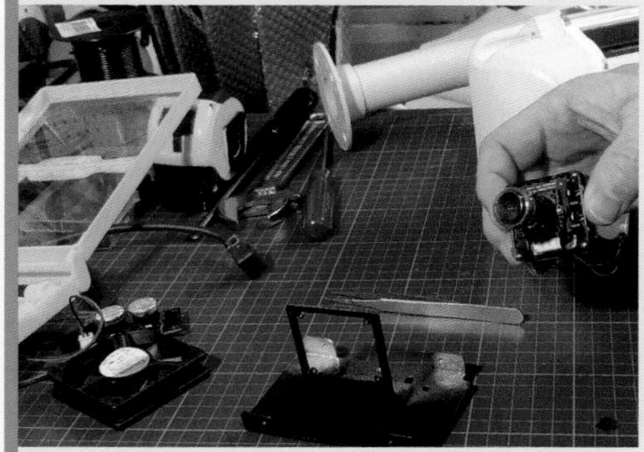

01 You will need a Raspberry Pi 3B+ or 4, HQ or equivalent camera, and a waterproof housing. Follow Jeremy's setup instructions provided at **magpi.cc/assemblingGMS**. Start by unscrewing the lens to remove the infrared filter, and cut off the camera mount nubbins so it sits flush.

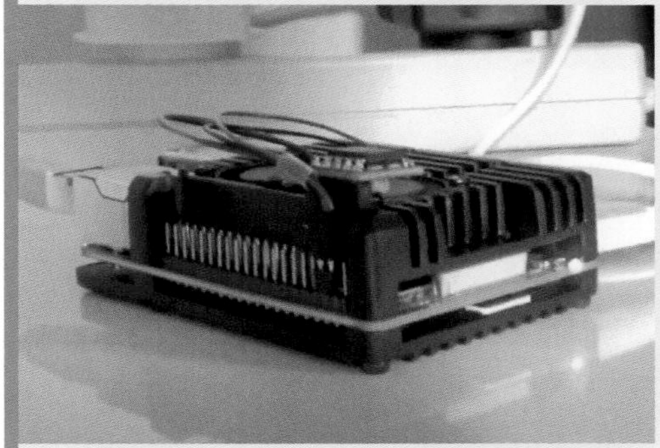

02 Attach a heatsink and run a power supply to Raspberry Pi, since you will need it to be running constantly if you want the best chance of capturing meteor footage.

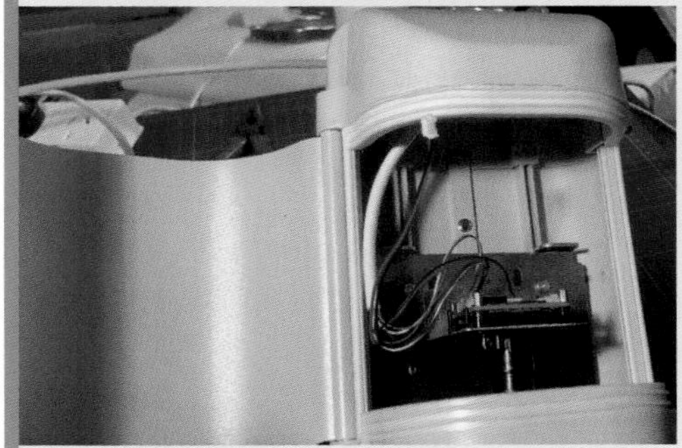

03 Connect everything together and place it inside a waterproof casing. Install somewhere there's a clear sky view, as with this Fireballs Aotearoa meteor camera at a school site in Otago, New Zealand.

LEGO
Submarine 4.0

The challenge of keeping a LEGO-based submarine afloat is solved with Raspberry Pi Zero W, as **Rosie Hattersley** finds out

MAKER

Brick Experiment Channel

Software engineer BEC has a lifelong passion for designing builds using LEGO Technics parts. Raspberry Pi has proved invaluable to their most recent projects.

magpi.cc/ brickexperiment

This LEGO builder extraordinaire is a "middle-aged guy from Finland" who dreams up, and then painstakingly creates, working models that address complex physics and mathematical challenges. His latest build, LEGO Submarine 4.0, runs off a LEGO EV3 motor and Raspberry Pi Zero 2 W. Like all his builds, the submarine project is thoroughly documented on his YouTube channel, Brick Experiment Channel (**magpi.cc/becyoutube**), where his methodical approach to construction and problem-solving have earned him more than 2.7 million subscribers.

This is the fourth LEGO submarine design he has completed. The first three variously used propellers to add or reduce buoyancy ("gravity and buoyancy stay always the same while the propellers exert

force"); a balloon, and an air compressor to adjust the amount of water displaced, thereby controlling whether the submarine sinks or rises; and a piston ballast to suck in more water to add weight and increase the sub's gravity.

He settled on the last method for Submarine 4.0, despite the difficulty of gauging the neutral buoyancy point. However, it had proved a more stable setup and would not compress under pressure when submerged. Most importantly, "you

▶ A submarine needs a captain, which was easily supplied in the form of LEGO part ID col154

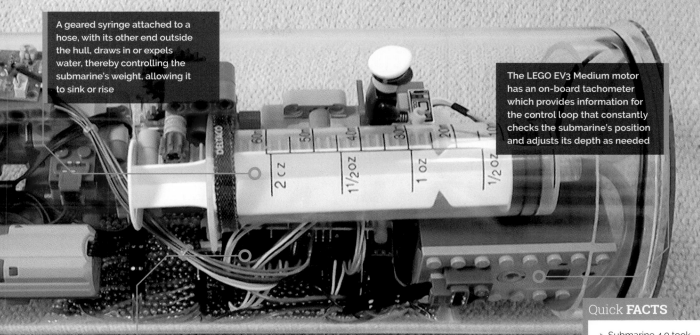

A geared syringe attached to a hose, with its other end outside the hull, draws in or expels water, thereby controlling the submarine's weight, allowing it to sink or rise

The LEGO EV3 Medium motor has an on-board tachometer which provides information for the control loop that constantly checks the submarine's position and adjusts its depth as needed

An absolute pressure sensor and laser sensor monitor the submarine's depth, while a Raspberry Pi Zero 2 W PID control loop uses this information to adjust the syringe ballast

Quick **FACTS**

> ▸ Submarine 4.0 took roughly 300 hours to make

> ▸ The maker's philosophy is that you learn by doing, one step at a time

> ▸ Code for the project is at: **magpi.cc/ sub4code**

> ▸ He enjoys the data logs that Raspberry Pi produces

> ▸ He uses this script to incorporate them in his videos: **magpi.cc/ log2labelpy**

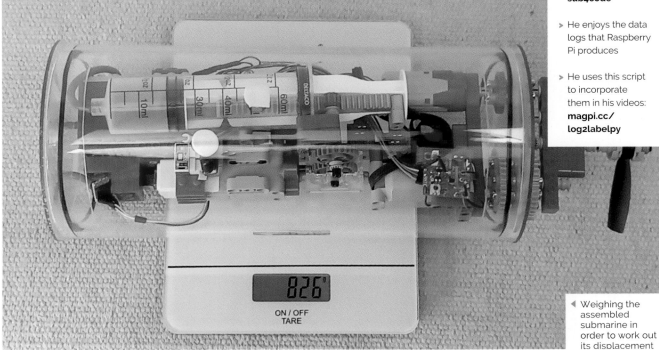

◀ Weighing the assembled submarine in order to work out its displacement

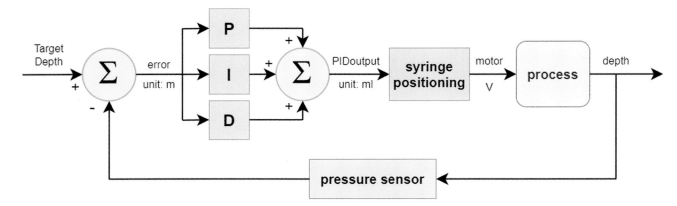

▲ Raspberry Pi's PID control loop takes depth information from the sensors and uses it to adjust the syringe ballast

❝ He attached a magnet on the inside top of the submarine frame so he could 'fish' the craft out of danger ❞

can measure the piston position with a LEGO EV3 motor that contains a tachometer. That will help the control loop." This loop is also the reason for using Raspberry Pi: as we reported last issue, he had recently made an impressive inverted pendulum that uses a PID (proportional integral derivative) control loop running on Raspberry Pi Zero 2 W to accurately measure and compensate for constantly changing speed, location, and pressure levels.

He planned a similar setup here, to monitor and control the submarine's depth. He soon found the wireless LAN connectivity invaluable when tweaking the PID parameters and updating any

▼ A rigorous series of underwater tests saw Submarine 4.0 submerged in a murky river

Python code without having to physically connect to the Pi Zero 2 W via USB (which would have involved carefully extracting everything from the precision environment he'd created). He now says wireless LAN is "an absolute necessity" which "made the development process a lot faster."

Underwater flying machine

A fair amount of time and effort went into creating the submarine's beautiful transparent acrylic case with tightly fitting and invisible end caps. The basic acrylic cylinders were precision-cut and end pieces with rubber seals attached to form a waterproof unit. LEGO gears were fitted to control the syringe that would adjust the buoyancy.

An absolute pressure sensor (which measures pressure relative to a vacuum, and is unaffected by the ambient pressure) is used to track the submarine's depth. It connects to Raspberry Pi Zero 2 W via I2C. A SparkFun TFMini-S Micro laser sensor provides a second means of measuring the submarine's depth, but its accuracy is affected by the murky environments in which he was using it.

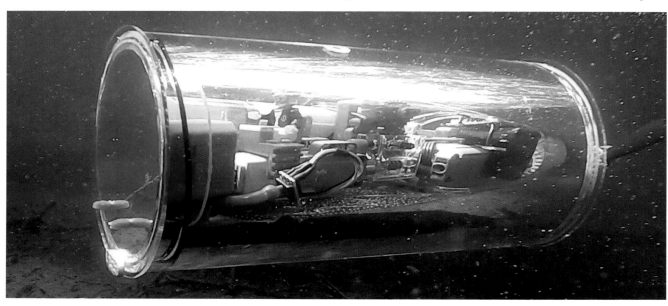

```
   time: 07:58
  depth: 15.2 cm
  laser: 7.9 cm
syringe: 27.2 ml
   temp: 25.0 °C
 button: forward
```

▲ River testing data with log info provided by Raspberry Pi

A more successful purchase was the radio board he harvested from a cheap Chinese toy submarine, having chosen it for its 27MHz radio frequency (needed to penetrate water) and its aesthetically pleasing controller. Raspberry Pi provides enough juice to power the board, so he decided to discard its LiPo battery in favour of a LEGO waterproof rechargeable battery pack.

Diving for pearls

In Submarines 2.0 and 3.0, he used lead pellets to provide extra weight, but they were quite sizable and took up valuable space inside the submarine's frame. For this version, he splashed out on expensive 2.5mm tungsten pellets weighing 18g/cm³. Weighed on a kitchen scale, the submarine was 826g, with a displacement of 1614g. He added 580g of tungsten pellets to make the submarine dive gently, making adjustments using the syringe.

To prevent entanglements while on manoeuvres, he attached a magnet on the inside top of the submarine frame so he could 'fish' the craft out of danger, if needs be. With lots of weeds and obstructions on the river bed, he was keen to avoid collisions, especially as it cost more than 600 EUR.

Thankfully, Submarine 4.0 has performed well in a range of environments from swimming pools and water tanks to a nearby river. "It drives well under water. The automatic depth control really makes controlling it easy, as you can focus on pressing only forward/backward and left/right buttons and forget the dive/surface buttons. I'd say the controls are as good as in Submarine 2.0, which has been the best so far," he reports.

Nonetheless, as a perfectionist, he notes several areas for improvement. ⚅

Underwater explorer

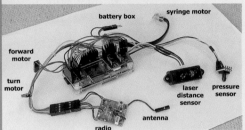

01 As well as a food syringe and LEGO gear wheels, you'll need a Raspberry Pi Zero 2 W with 16GB microSD card, a LEGO EV3 Medium motor with tachometer, and a 27MHz radio controller.

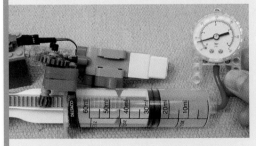

02 A motorised LEGO gearing system adjusts the ballast syringe. A hose sucks in water and draws water into the syringe chamber. Buoyancy remains constant while the extra weight of water increases the gravitational pressure.

03 A radio board and controller culled from a cheap toy provide a user-friendly remote control, powered by Raspberry Pi, with no need for a dedicated battery.

PoleFX

The networking power of Raspberry Pi drives LED lights to liven up acrobatic dance routines. **Lucy Hattersley** gives PoleFX a twirl

MAKER

Spencer Hochberg

Spencer spent his teenage years inventing unicycle tricks with photography and videography as a side effect. He earned a degree in mechanical engineering from UCLA, and now works with small teams on live shows.

polefx.com

Billing itself as the world's most advanced dance pole, PoleFX uses thousands of LED lights embedded inside a network-connected pole. On the sidelines, a Raspberry Pi-powered network box is using OpenCV to image match videos and turn them into light-based visual animations.

"Our goal is to create functional pieces of art and embed show technology into new places," explains Spencer Hochberg, inventor and owner of PoleFX.

Amongst the dazzle and display, it is easy to overlook the technical requirements of this build. The pole itself needs to be "structurally strong" explains Spencer, to withstand the dynamic moves of acrobats, "while protecting and displaying an array of thousands of integrated pixels."

Those thousands of pixels create an LED screen inside the pole structure. "The pole acts as a canvas for digital art that complements the performer," explains Spencer.

Networked light

Inside PoleFX sits a Raspberry Pi acting as the system's content server and control interface. Each pole is connected to PoleFX via an Ethernet cable, and the poles receive frames of data "on-the-fly" from Raspberry Pi.

The control surface is a local web interface. "We also support simple control using a USB keyboard, encoder with OLED display, as well as DMX [Digital Multiplex lighting controller]," Spencer tells us.

"Raspberry Pi made it easy for us to put a flexible, low-cost brain in the middle of our system. The giant community of users has made it easy to find tutorials to get started and configure everything," he notes.

The low power requirements for Raspberry Pi were also a consideration, and the "huge amount of online resources" made it quick and painless to implement.

The current poles are powered using a Raspberry Pi 3 Model B, which is processing and sending frames out at 60 fps.

Pole design

The poles had to be specially designed and custom-built for purpose. The outer shell of the pole is made using "extruded black polycarbonate tubes." Spencer tells us this material is "resistant to cracking and hides the LEDs." They allow the light to shine through.

Inside the pole, 3D-printed parts are used to mount non-structural components, and an internal slip ring that enables the wire connections to spin as the pole rotates.

PoleFX is "mostly built from off the shelf components." The main mounting plate inside is a custom PCB (printed circuit board), which is an "easy way to fabricate a mounting plate and simplifies the internal wiring." The back panel of the enclosure where the ports are exposed is also a PCB, used only for mechanical and cosmetic qualities.

Smart software

The main code is written in Python and the patterns are encoded as H.264 video files. The video frames are read using OpenCV to sample the pixel values. "We then composite and make other post-processing adjustments before sending the pixel data out over the Ethernet network in multicast sACN/E1.31 packets," Spencer reveals.

The outer shell of the pole is made using extruded black polycarbonate tubes that allow light to shine through

Inside the pole are thousands of LED lights with the pattern determined by OpenCV scanning a video file

An internal slip ring enables the wire connections to spin as the pole rotates

Quick **FACTS**

- One of Spencer's first projects was an LED unicycle wheel

- Spencer designed the first LED pole prototype six years ago

- It started as a collaboration with performance company SpinFX

- The free-standing and aerial poles cost just $100

- The poles are approved for a maximum weight of 330 lbs (150 kg)

▲ The PoleFX box contains a Raspberry Pi and the ESP32 board, and connects to the poles via Ethernet cables

▶ The animated poles enable the dance to have an extra layer of expression while the dancers perform their routines

◀ The poles need to be structurally strong to withstand the acrobatic moves while protecting/ displaying the array of thousands of LEDs

How PoleFX works

01 Inside the PoleFX box is a custom ESP32-based board with an Ethernet interface. Each pole is connected to the board via an Ethernet cable which transmits the data to the LEDs.

Raspberry Pi is configured as a wireless LAN hotspot that hosts an Open Stage Control server, so the user can connect to the network and load the control UI without having to install anything.

Apart from controlling dance poles, the content-server side of the project is generally useful for storing, generating, and playing back

▲ The pattern is created by scanning video files with image recognition software

02 A local web interface is used to select different patterns and add text to the pole design.

> ## Touch-reactive LED handrails and jungle gyms may be just around the corner

LED animations for a wide range of installations where a whole PC would be "overkill", and "a microcontroller alone is too limited."

An early motivation for building load-bearing, pixel-mapped poles was then use them to build larger structures. "Touch-reactive LED handrails and jungle gyms may be just around the corner!" ▥

03 The pattern on the pole is created using OpenCV to pattern-match a H.264 video stream (which is converted to a matching LED display).

Cyberdog
Smart Saddle

Smart pups sport luminous collars, but true fashionistas shimmer in NeoPixels, courtesy of Raspberry Pi. **Rosie Hattersley** reports from the canine catwalk

MAKER

Kevin McAleer

Kevin makes robots, brings them to life with code, and makes videos about them on YouTube.

magpi.cc/ kevsrobots

Warning!
Heat

Without being alarmist, strapping your pet into a jacket clad in electrical currents could cause discomfort if the circuit produces too much heat. Choose a power supply and controller with current limited to prevent overheating. Be sure to check the heat levels of the jacket before fitting.

magpi.cc/ currentlimiting

With the change of seasons, self-confessed "Raspberry Pi nerd" Kevin McAleer figured his beloved pups could do with some eye-catching outerwear to ensure they'd be seen during their evening walk. After a quick consultation with his chihuahua, Archie, about his modelling fees, Kevin set about designing the NeoPixel-encrusted Cyberdog Smart Saddle, an RGB LED coat that Kevin would be able to control from his phone.

Brighter, later

Kevin McAleer's work may be familiar: he hosts a popular Sunday YouTube robot design channel (**magpi.cc/kevsrobots**). Among other intriguing makes, we've recently featured both Kevin's Billy Bass remotely flappable fish (**magpi.cc/billybass**) and his PIKON DIY camera (**magpi.cc/pikoncamera**). While both those projects upcycle existing hardware, the idea behind the delightful doggie Day-Glo jacket was "to create something fun from some foot-long LED NeoPixel strips." Kevin realised that he could control these from a couple of feet away using a Pimoroni Plasma 2040 microcontroller.

The Plasma 2040 also has a very important current-limiting function that ensures Archie doesn't end up toasted, should the lights short out and draw a lot of current very quickly. However, the Plasma lacks both Bluetooth and wireless connectivity. This is where Raspberry Pi Pico W comes in. With the low cost of Pico Ws,

▲ Admiring looks while sashaying along Blackpool promenade

> Assembly and testing took two evenings

> Two coats of superglue were needed…

> …"in order to withstand a dog's rigorous shake"

> Kevin's parts list includes a £2 PVC dog coat…

> …and a cute chihuahua: "Priceless!"

Where better to show off some fancy illuminations than Blackpool? The light show is controlled via smartphone, with Pico W providing connectivity

The 3D-printed saddle contains a tiny LiPo battery, Raspberry Pi Pico W, and Plasma 2040

A Pimoroni Plasma 2040 LED strip driver is specifically designed to power multiple NeoPixel lights, crucially without getting overly warm

▲ Pimoroni Plasma 2040 and Raspberry Pi Pico W jointly operate the rainbow smart saddle

▲ Spacing the LED strips correctly to work with scrolling text was a challenge

Kevin thought, "well, why not just add one to the project and have it act as a bridge and front-end to the project? It can handle all the wireless stuff, whereas the Plasma 2040 can do what it's best at – making LEDs light up."

Kevin also had a cunning plan for powering everything: he'd run it all off a tiny LiPo battery, secreted in a self-contained harness that can comfortably sit on Archie's back. Fingers crossed, the battery would last long enough for a stroll along Blackpool promenade, the home of illuminations and the perfect place to show off such a marvellously attired pup.

Style and practicality

Designing the 3D-printed saddle was probably the biggest challenge, since it had to comfortably fit Archie while also holding all the electronics. "Will it fit right? Can it safely hold all the components? Will it be light enough?" were Kevin's chief concerns here. He also worried whether the attention-grabbing effect would last long enough for a reasonable length stroll, and didn't want Archie to have to style out a wardrobe malfunction. Adding a pocket to the saddle design to allow for speedy battery changes helped allay this fear, while applying lashings of superglue helped adhere the rainbow LED strips to the plastic dog coat.

Given his choice of microcontrollers, this is a MicroPython-based project, which suited Kevin very well. "I love MicroPython," he declares. "All

01 Use pencil and paper to measure up your pup for their Cyberdog saddle, taking their contours into account. Add a tiny pocket to your saddle design: it's ideal for keeping a LiPo battery that can be easily replaced.

02 Connect the battery to connect to a Pimoroni LiPo Amigo Pro. This will charge it and can also connect to another device via a JST-PH connector. Kevin recommends Pimoroni's Phew! Code to add a wireless access point.

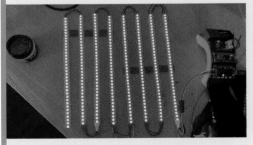

03 Check that the setup works and the lights illuminate as expected. 'Dog-proof' the LED strips by applying two or more lots of superglue, so there's less chance of them coming unstuck from the plastic dog coat.

the code [for the project] is written in this for simplicity and speed of creation. It's as close to English as any language I'm aware of; good code reads like regular sentences."

▲ Even single-colour LEDs attract attention, let alone Archie's full light show

Kevin credits the "amazing" Phew! code from Pimoroni (**magpi.cc/phewgit**) for the easy setup

❝ Designing the 3D-printed saddle was probably the biggest challenge ❞

of wireless access point and web interface. Phew! (Pico HTTP Endpoint Wrangler) "does a couple of cool things," Kevin explains. "It can make the Pico W into an access point for a device, meaning it will appear on your phone as a wireless hotspot."

Another handy feature meant Kevin was able to add scrolling text as a final flourish to his already impressive illuminated Cyberdog Smart Saddle. Since Phew! can divert traffic to a specific web page, he was able to use it to change the LED patterns or scroll text he entered, and have it appear on the Cyberdog Smart Saddle. Cheeky!

"I'm not aware of any project that has influenced this," says Kevin. "It's just a crazy idea I saw through to completion!" We think it's doggone good! ⩗

Arribada Penguin Monitoring

Getting closer to nature to understand creatures' lives is perhaps the perfect use of a Raspberry Pi and time-lapse camera pairing, thinks **Rosie Hattersley**

MAKER

Alasdair Davies

Conservation scientist and NatureBytes co-founder Alasdair set up Arribada to provide low-cost, open-source wildlife monitoring hardware.

arribada.org

Cute creatures rule the internet and social media, which is absolutely as it should be. So it follows that photos of otherwise hard-to-encounter, impossibly personable penguins going about their daily business in the depths of an Antarctic winter, are likely to be pretty popular. The challenge, of course, is getting to the extreme south in the first place, and setting up a camera that can snap such shots and withstand temperatures as low as –30°C. Arribada's Alasdair Davies decided he and his Raspberry Pi–based camera kit were up to the challenge of a long-term time-lapse photography project. Penguin Watch (**magpi.cc/penguinwatch**) takes and posts photos to a website, where citizen scientists avidly spot penguins. This enables the enterprise to monitor population numbers. What none of them could anticipate was the three-year gap between setting up the new cameras and being able to retrieve the resulting photos.

Sea change scenario

Alasdair's wildlife monitoring and conservation work at The Zoological Society of London (ZSL) resulted in a Shuttleworth fellowship where he designed and built "anything from a special camera trap that would wake up and detect an

> ❝ Their Raspberry Pi time-lapse camera setup had worked like a dream! ❞

animal walking by, to a device that would send back an alert if a particular animal, such as an elephant, turned up." Alasdair founded Arribada in 2017, and explains that his key aspiration was to get into open-source conservation technology so he could start to share some of the designs and make it more accessible for others to get involved.

He wanted to be able to put wildlife monitoring kits into the hands of young citizen scientists, have them interact with the live data and report back on changes over time. Surprise, surprise,

A Raspberry Pi 3B and camera operate from within the ruggedised case, capturing photos hourly and saving them to an SD card

Battery power and wireless connectivity are provided by the solar panel. A Python script initiates the hourly photo, then everything goes back into standby

Observing penguin behaviour and population numbers over time

◀ To preserve power, few photos are scheduled during the darker months, when penguins are, in any case, much less active

Quick **FACTS**

➤ Remote monitoring involves hitch-hiking on research vessels...

➤ ...and dropping off/retrieving time-lapse cameras as lifts allow

➤ Far too many time-lapse photos are generated each year

➤ But citizen scientists can help analyse them...

➤ ...by counting the penguins in each photo

Raspberry Pi loomed large in these plans. "Raspberry Pi had always been one of the tools I use because it was so accessible and affordable for anyone to get involved. And I was using it myself in a lot of the kind of products I was making at the time." The name 'arribada' means arrival, and is specifically associated with the migration and birth cycles of sea turtles. In fact, one of Arribada's earliest projects was creating a Raspberry Pi Zero-based device in a watertight shell that could be attached to a sea turtle and video its movements.

▼ In another Arribada project off Principe, a turtle equipped with a Raspberry Pi Zero-powered video camera on its back spies a friend

Creating a suitable device to monitor penguin populations in Antarctica involved an even more challenging environment. Soon enough, a penguinologist (yes, really!) from Penguin Watch got in touch about low-cost wildlife monitoring cameras that could provide evidence of long-term changes to its subjects' behaviour and habitat, and form part of the conversation about climate policies. "They wanted to lower the cost of taking time-lapse photography of penguin colonies, watch the penguin colonies throughout the seasons, to understand if the penguin colony was affected by changes such as whether the lack of sea ice was causing more predation, and also whether it was delaying their nesting and feeding." These are all questions you can answer if you have a camera and can point it at a colony, he points out.

Alasdair needed to come up with a camera setup resilient enough to withstand Antarctic winters and still deliver a regular stream of photos showing penguins' everyday lives. Cost reduction was a huge factor: using Raspberry Pi Zero and a camera, along with a PiRA Zero power scheduling unit and solar panels, has cut costs by two-thirds to around £100, and is so energy-efficient that the setup can run for several years at a time.

The first time-lapse camera was programmed to wake up and take a photo once an hour every hour,

DIY critter watch

01 Armed with a fresh installation of Raspberry Pi OS, a Pi HQ Camera, and a generous capacity SD card, follow the instructions at **magpi.cc/arribadamonitor** to set up time-lapse photos. Their example is for photographing orchids.

> **They wanted to lower the cost of taking time-lapse photography of penguin colonies**

▲ Successive images taken from the same location show the changing habitat

02 If the setup is to be used remotely, use a PiRA Zero (**irnas.eu/pira**) to schedule things, while a small solar panel might come in useful to ensure ongoing power.

and was installed in February 2019 after hitching a lift on one of many "ships of opportunity" that visit Antarctica during its summer. There was no connectivity and no such opportunity to retrieve the camera from the peninsula, as planned the next year, due to the events of 2020. In 2021, "that particular day when the boat was going past our location, sea ice blocked entry so there was no way of getting to land and no chance of retrieving the camera," Alasdair relates. The camera was finally retrieved in early 2022 when the team had the first chance to see whether their penguin monitoring had worked. Alasdair received the package in the post. "It stunk of penguin!" He placed the SD card into his computer: "it was just reading and reading for quite some time and suddenly stopped. And it said it discovered something like 32,764 items." They'd got 32,000 photos of penguins and their habitat. Their Raspberry Pi time-lapse camera setup had worked like a dream! 🐧

03 Most importantly, encase your hardware in a waterproof case and site your time-lapse camera out of harm's way, perhaps securely lashed to a pole or tree, angled to where you know your critter often hangs out.

BMOctoprint

Building an OctoPrint server for your 3D printer is cool, but building it
into a cute gaming robot body is much cooler, as **Rob Zwetsloot** finds out

MAKER

Allie Katz

Artist, technologist,
maker, and
YouTuber. Allie
shoots videos
of their creative
adventures in
a wide array of
builder hobbies.

**magpi.cc/
katzcreates**

f you've ever seen *Adventure Time*, the strange
and cool cartoon that started about twelve
years ago, you might remember the sentient
gaming robot and friend of the main characters,
BMO. This cute retro-inspired, handheld robot
has been made many times with Raspberry Pi
to be an actual games console or interactive
costume. However, Allie Katz, of Katz Creates, did
something a little different.

"BMOctoprint is the portmanteau of BMO from
Adventure Time, and OctoPrint, the 3D printing

server for remote access, and that is exactly what
he is," Allie explains to us. "BMOctoprint is a
life-sized replica of the beloved cartoon robot,
powered by a Raspberry Pi 4, in a fully 3D-printed
case, with functional buttons and a touchscreen
face. He can basically control your 3D printer,
whilst also giving you compliments and making
you laugh, which officially makes him the best 3D
printer server ever."

While looking for a case for their Raspberry
Pi 4, a patron suggested to Allie that they could
put it inside a BMO-shaped case. Allie loved the
idea as they're a fan of BMO, who is a bit of a
non-binary icon.

Long time coming

"The build process – and, moreover, the design
process – was extensive and rigorous for this
because I knew from the get-go that I wanted
to share the final plans online so others could
make their own BMOctoprint," Allie tells us.
"Designing and building something for yourself,
that will be done once and never again, is a walk
in the park compared to creating something
that will be touched by the hands of countless
others in a way that you cannot ever fully
anticipate. User experience and accessibility are
of paramount importance to me and, as such, I
wanted BMOctoprint to be fun and easy to build,
as much as possible."

The designs are made so they can be 3D-printed
as easily as possible, and the whole thing has been
documented on the Katz Creates YouTube channel.

With BMOctoprint connected to a 3D printer
with a USB port, on boot it shows the OctoPrint
GUI which can be controlled with the touchscreen.

"A lovely little idle animation works as a
screensaver, meaning whenever you aren't actively

▲ All the electronics neatly fit inside BMO,
leaving room to access and maintain as well

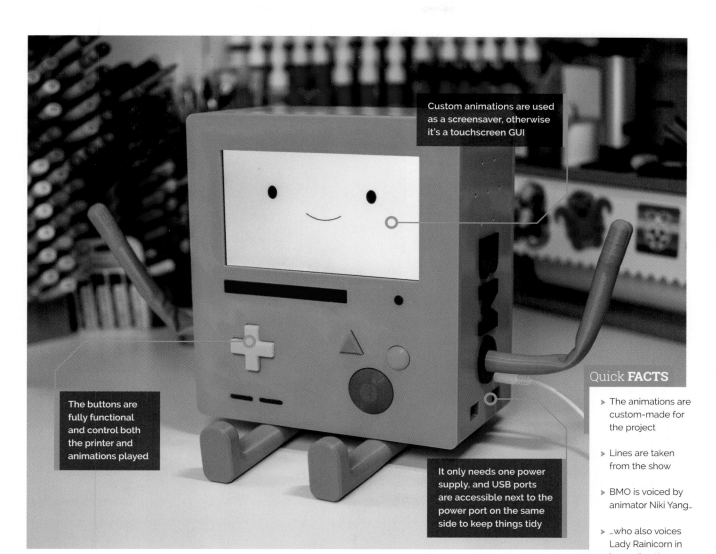

Custom animations are used as a screensaver, otherwise it's a touchscreen GUI

The buttons are fully functional and control both the printer and animations played

It only needs one power supply, and USB ports are accessible next to the power port on the same side to keep things tidy

Quick **FACTS**

> The animations are custom-made for the project

> Lines are taken from the show

> BMO is voiced by animator Niki Yang…

> …who also voices Lady Rainicorn in her native Korean

> All the files and code can be found here: **magpi.cc/ bmoctoprint**

> User experience and accessibility are of paramount importance to me and, as such, I wanted BMOctoprint to be fun and easy to build

▲ Some work needs to be done on the prints to make them ready

using the touchscreen, BMO wakes up and starts looking around," Allie says. "The buttons on the right of the faceplate control the 3D printer and trigger a range of commands, from connecting to the printer to cancelling and pausing prints. The D-pad buttons are functional as well, but instead of controlling the printer, they each trigger a unique custom-animated reaction from BMO."

▲ The electronics are fairly simple, even using a custom PCB

▶ The buttons are fully functional, with the D-pad activating specific animations

▲ I/O ports are kept on one side so that everything is neat and easily accessible

Distant future

We love the very clean and accurate look of BMOctoprint, and so do a lot of people in the wider maker community, if social media reactions have been anything to go by. While Allie hopes people will make their own, they're not done yet.

"I definitely want to add a Raspberry Pi Camera Module for watching prints, and have already accommodated this by adding a small slot in the back plate where the ribbon cable can fit through," Allie explains. "I also think BMOctoprint should have things he can hold, since there are magnets already places in the hands, though I have yet to come up with exactly what! I'm also pondering using a USB microphone to try adding voice activation into the mix, but that's still a wish list item at the moment!" ◪

Making something for everyone

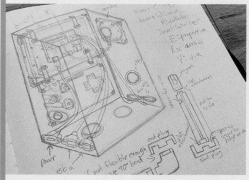

01 Designing something that can be used consistently and easily replicated is very different from a one-off project. Allie spent ages on the design for BMOctoprint as a result.

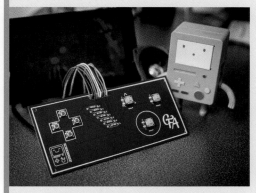

02 Thoroughly testing the electronics is an important part of any build process, especially when using custom PCBs like Allie made.

03 Making sure it can be constructed easily is also very important – sometimes you need to check the tolerances, especially with a 3D print.

ForestryPi

A professional arborist used Raspberry Pi hemispheric monitoring to
observe trees across several seasons. **Rosie Hattersley** hears the story

MAKER

Edward Lane

Ed Lane learned
about low-cost
nature monitoring
while studying and
worked backward
to create his own,
albeit with "just
disgusting" code.

magpi.cc/
hemiphoto

With an established tree surgery business under their belts, Ed Lane and his brother decided to branch out into "proper forestry", offering advice to larger clients such as farmers and landowners about the long-term health of their land. Obvious
puns aside, adding hemispheric photographic monitoring to their list of services meant both clients and arborists could take a long-term look at how climate change was impacting stands of trees in their immediate vicinity. Having encountered Raspberry Pi at university, and through a general interest in technology, Ed chose it for Green Lane Forestry's ForestryPi monitoring scheme, which quickly attracted several business customers.

In a further twist, Ed has recently handed over the tree surgery business to his brother and is currently training as a primary school teacher. He has taken one of his ForestryPi monitoring kits with him and will use it to inspire pupils as citizen scientists, reporting on the trees at school over the seasons and successive years.

> ❛❛ He has taken one of his ForestryPi monitoring kits with him ❜❜

International inspiration
Ed returned to university to study aquaculture and fisheries, where he was struck by the number of examples of academic papers detailing "really cheap bits of computer equipment to monitor XYZ" in poor parts of the world. "I was always really interested in those papers, and how little, cheap computers really can democratise science and give everyone the chance to do fairly complicated stuff that would have, a few years ago, required thousands of pounds worth of kit." Having read a paper last year (**magpi.cc/canopydynamics**) about monitoring a forest canopy, Ed decided he could do something similar, despite being "no expert coder." A Raspberry Pi Zero, Witty Pi 3 real-time

▶ Weeding out the
best photos, and
using ImageJ to
process them,
produces some
great results

The ForestryPi box is battery-powered and uses a Witty Pi 3 real-time clock to ensure photos are taken at preset intervals

The setup is ideal for long-term monitoring of whole stands of trees across multiple seasons

Raspberry Pi Zero and a fish-eye camera take photos of the leaf coverage of a tree on agricultural land near Arundel, Sussex

Quick **FACTS**

➤ Ed says he is "no expert coder"

➤ A friend called his code "just disgusting" – but it works!

➤ Ed's pupils will benefit from ForestryPi

➤ He's set it up at the primary school where he's based

➤ It will become a multi-year citizen science project

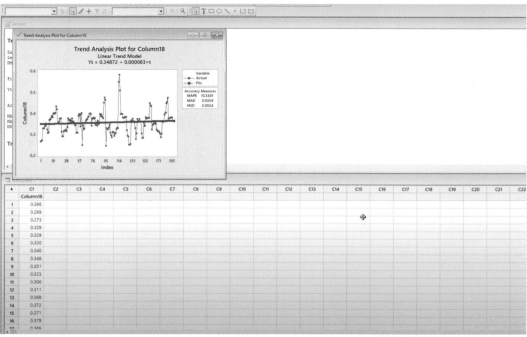

◀ Adding the images to a database helps track leaf coverage over time

Take hemispheric photos

01 ForestryPi is based around a Raspberry Pi Zero with a 32GB SD card, so there is plenty of storage space for the photos it takes over a lengthy period. Find setup code and instructions at **magpi.cc/forestcamtut**.

02 Attach a fish-eye lens to Raspberry Pi and add cables for a powerful battery pack so the setup can run without user intervention for several months.

03 Attach a Witty Pi 3 real-time clock so the Raspberry Pi knows when to activate the camera shutter. House everything inside a sturdy, weatherproof box and attach securely to a tree trunk, with the camera pointing upwards to capture leaf canopy changes.

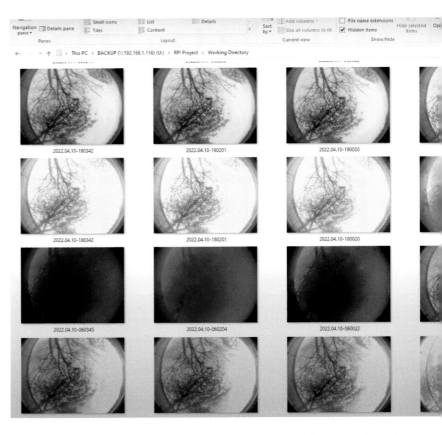

▲ Photos downloaded from ForestryPi's SD card show the variable image quality involved in automated photography

clock, fish-eye lens, 20,000 mAh battery pack, and a 32GB SD card, plus a plastic casing, formed the basis of the ForestryPi setup.

Ed details how he set up ForestryPi on the Green Lane Forestry website (**magpi.cc/hemiguide**), largely using software such as Microsoft RStudio he had used while studying, along with ImageJ (**imagej.net**) to process the photos the time-lapse

> ❝ His approach of working backwards to get something serviceable nonetheless bore fruit ❞

camera produced. The camera itself is mounted on the trunk of a tree in a weatherproof box. Keeping the rain out was a big challenge for Ed, who came up with the parts list and built the complete ForestryPi monitoring system himself. A local land agent expressed an interest in the project and they settled on a suitable stand of trees, planning to "generate some data to see the gradual growth and decline of the tree canopy over the course of the growing season and into winter." A computer scientist friend was "just disgusted" at the code Ed

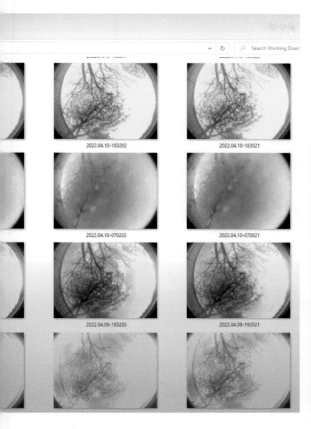

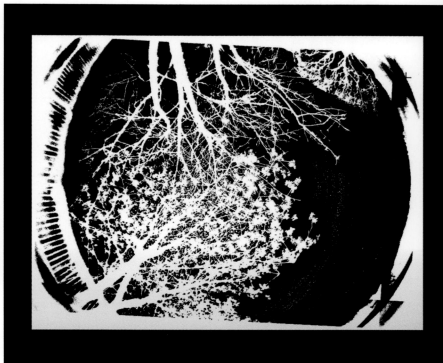

produced, but his approach of working backwards to get something serviceable nonetheless bore fruit.

Rewarding results

ForestryPi is set to take photographs at regular intervals, but the changeable weather conditions in West Sussex, where Ed is based, mean some photos are far more useful than others. Ed says there were "all sorts of problems of light and shadow. I thought it would work fine regardless but, actually, when you come to analyse the image you realise, 'Nope, that's too early in the day'." He also found issues with "weird artifacting" and a lot of scattering of data, along with some interesting results. In fact, Ed sees lots of potential for the setup, especially if a camera can be set up to record the life of a tree or stand of trees over multiple years.

"For the best results, you really do need a woodland where you've got cover on all sides," says Ed. Few people have ready access to a stand of trees whose canopy you can monitor, but lots of people have trees in their gardens. Ed adds that "anyone with an interest in Raspberry Pi is capable of giving this a go and monitoring the growth of their own trees. It's a fun project, and it provides a good introduction to using Raspberry Pi in a remote setting." ◪

▲ Witty Pi 3 is used to add a real-time clock and power management to the project

The Lost Sounds Project

A noticeable reduction in birdsong led a digital technologist to set up a Raspberry Pi-based nature-watching project for schoolchildren, learns **Rosie Hattersley**

MAKER

Dr Liz Edwards

Liz is an interdisciplinary researcher working in the Future Places Centre at Lancaster University, where she uses digital technologies and code in applied projects.

thelostsounds.org

I dentifying creatures based on their appearance is a matter for the observant, with a little help from reference books, related websites and apps, and maybe some clever AI.

Sound identification requires a different set of tools, and is reliant on the bird or animal in question first having been recorded and positively identified so that subsequent naturalists may be sure of what they're hearing. The Lost Sounds Project puts this recording process in the hands of schoolchildren who explore their local environment and use Raspberry Pi computers and microphones to record bird songs, as well as their unique characteristics.

Mixed media

The Lost Sounds Project was piloted in Morecambe Bay, "one of the most significant sites in the UK for breeding birds," explains Dr Liz Edwards, part of the Ensemble team at Lancaster University's School of Computing & Communications. Liz, herself, has a PhD in digital technologies to interpret public gardens, and experience both as a geography teacher and in multimedia design. "I'm not a computer scientist, but I use digital technologies and code in applied projects to do with nature connection and environmental science," she says.

Inspired by Scott Garner's wonderful 'Beet Box' (**magpi.cc/beetbox**), Liz went on to create the equally out there Rhubaphone, which invited

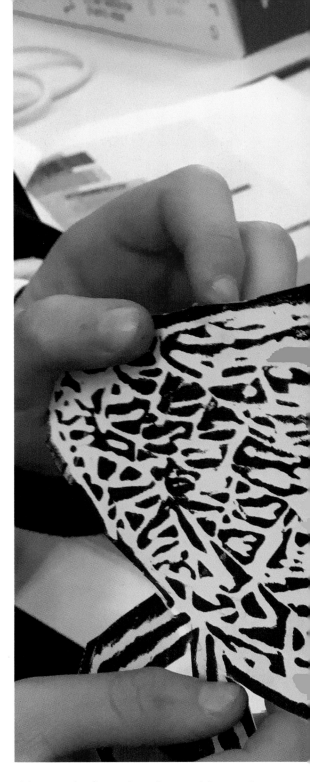

visitors to Clumber Park gardens to pick up various cultivars and listen to details of their differences and quirks (see **magpi.cc/talkingrhubarb**). Both such projects, of course, use Raspberry Pi.

Lost Sounds, meanwhile, takes its name and inspiration from "the power of *The Lost Words* book by Robert Macfarlane and Jackie Morris, and by several conversations noting the changing soundscape of the area, including the loss of corncrakes and the decline of curlew." The nature study project is often run as a cross-curricular activity in primary schools (for which Liz has

The bird image is clipped to a circuit. When someone presses on the bird print, the circuit is complete and the appropriate bird sound plays

Having identified the bird sound they heard, participants select its preprinted fabric image, which has been printed with conductive ink

Pupils and visitor centre participants use a pi-top, with a Raspberry Pi inside, to code instructions in Scratch

▶ Pressing down on the fabric bird print creates a circuit and makes the birdsong play

▶ An interactive soundscape of Morecambe Bay created by pupils at West End Primary School in Morecambe

▼ Raspberry Pi 3 is used to create bird sounds

put together resources and runs teacher training sessions), but you might also spot it at beacon sites such as RSPB nature reserves, the Eden Project, and so on.

Noticing and responding to the local environment is the common thread. Liz lists birdwatching, listening to bird calls with directional microphones, making conductive ink prints, learning to 'sing' bird calls, looking at sonogram shapes of bird calls, and creating interactive displays to share with school or community groups. The more tech-focused activities involve making a circuit with a capacitance sensor, then writing code in Scratch so that the correct bird's call plays when the corresponding image primed with capacitive ink is touched. An example of this can be seen on Vimeo (**magpi.cc/birdcallprints**).

The sounds are recorded locally, but can also be cross-referenced against an open-source bird song directory such as xeno-canto (**xeno-canto.org**).

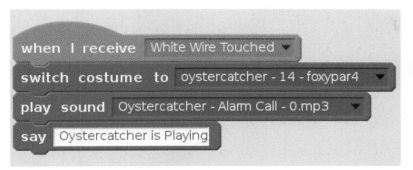

```
when I receive  White Wire Touched ▾

switch costume to  oystercatcher - 14 - foxypar4 ▾

play sound  Oystercatcher - Alarm Call - 0.mp3 ▾

say  Oystercatcher is Playing
```

Feathered friends

Liz came to use Raspberry Pi for The Lost Sounds
having tried several technologies and coding
environments while prototyping the project.
"My colleague (and friend of *The MagPi*) Lorraine
Underwood was incredibly helpful in this initial
development stage," says Liz. Raspberry Pi had
several standout features: "the ability to store
substantial image and audio files, the ability to run
Scratch, and incorporate circuit making."

Pi-tops became part of the mix when Liz
realised it was far more straightforward to provide
a consistent set of computing kit than deal with

▲ Working in pairs or
threes, children use
Scratch code to instruct
when each sound
should play

> 🔖 Noticing and responding
> to the local environment
> is the common thread 🔖

different IT setups at schools with whom she
was collaborating. She also uses Bare Conductive
Touch Boards for the large interactive soundscape
installations that are left in the venue after the
workshop. "It is an easy system for external
people to manage independently," she reasons. 🅼

◀ Children follow Scratch
code instructions to play
the correct bird sound
when they spot it

Interactive birdsong

01 Record and identify bird sounds in your area
using a directional microphone, checking the
species against a database such as **xeno-canto.org**.
Print images of the birds identified. Liz recommends
printing with conductive ink, if available.

02 Create a simple circuit with a capacitive sensor
connected to a Raspberry Pi. If you used
conductive ink for your bird print, you can add it to
the circuit using a crocodile clip. Other connection
options are possible if you didn't use such ink.

03 Use Scratch to create code to display a photo
of each bird species and play its song when the
circuit is complete.

K-9

Recreating a beloved TV character or prop is fairly exciting. **Rosie Hattersley** hears how Raspberry Pi allows one maker to give K-9 multiple Whovian features

MAKER

Fitz Walker

Fitz is a maker/tinkerer by nature with an unhealthy love for radio-controlled models of all types.

magpi.cc/hobbyview

Like many millions of us, "maker/tinkerer" Fitz Walker holds classic *Doctor Who* in very high regard. As an electrical engineer and computer scientist, he was better placed than many Whovians to show off his love of the show by building his own replica K-9, giving it "modern components" with Raspberry Pi.

Keeping the 'use whatever you've got to hand' ethos of the original *Doctor Who* sets, a loo roll holder disguises an unexpected K-9 defence mechanism.

The K-9 project helped Fitz fulfil an ambition to build a film replica of one of their "favourite companions to my favourite Doctor" [Tom Baker, naturally] and "to experiment with Raspberry Pi and other microcontrollers, and electronics to see how well they might replicate all the animatronic and visual features of the hero prop."

Fitz chose Raspberry Pi for its "powerful computer processing and video display", and options for interfacing with the physical world, giving K-9 "some level of autonomy and AI."

Design cues

Having been introduced to Raspberry Pi by a friend – who he thought was pranking him, based on the single-board computer's name – Fitz realised it would suit his needs very nicely. "Once I realised its potential, I kept Raspberry Pi in mind for any projects I engaged in." Raspberry Pi is "a powerful tool to do pretty much anything I can think of."

K-9 is the main focus of Fitz's tinkering efforts, and the only project for which he has had to do any custom design work, since some aspects could be adapted from other people's projects. In particular, founder of the K-9 Replica group Dave Everett had produced detailed templates and assembly guides for the animatronic dog's body, including a basic mechanical framework

The Tom Baker scarf is a nod to Fitz's favourite Doctor. He makes regular appearances on K-9's built-in screen displaying old TV clips

A Raspberry Pi 3 currently provides K-9's 'brains', helping him make tactical decisions and interact with humans and aliens

(**magpi.cc/k9bodyconstruction**), which were quickly seized on by fellow Whovian makers. Another impressive K-9 Raspberry Pi build that we've covered (**magpi.cc/k9prop**) uses this template too, or you can join a dedicated builder's Facebook group: **magpi.cc/k9facebook**.

Building on the original

Fitz also made use of existing designs for the very intriguing camera and AI element that he added to K-9's head. This is "based, in part, on the wedding photo booth (**magpi.cc/photoboothdiy**) by Anthony

K-9's head contains a camera. Photos are analysed by Raspberry Pi before K-9 prints out a pronouncement on whether a human or alien is in sight

Quick **FACTS**

> Fitz's K-9 extras include a loo roll holder nose…

> …which conceals a retractable laser gun

> There's a side monitor originally designed for cars…

> …which plays a random selection of *Dr Who* clips…

> …or sometimes random waveform sounds from YouTube

Sabatella which inspired K-9's photo-taking abilities," code which Fitz modified for K-9.

"Pretty much all the electronics and most of the animatronics were designed and set up by myself," says Fitz, who reports that the build was not complex but required "a bit of time and patience to build, program, and debug," with integrating all the parts the most challenging aspect. "Like any project, unexpected things come up. The drive system was a bit troublesome at first and had to be totally redone. Even then, it has some directional instability that needed to be rectified with a gyro."

▲ Elementary school pupils have been lucky enough to get a surprise visit too

K-9 Sys
Power Level:
57 %
Voltage Level:
1139
Error Status:
None

POLICE PUBLIC CALL BOX

TO VICTORY!

▲ Fitz's K-9 has made various appearances at shows in the US, often flanked by Daleks

K-9 is a combination of 3D-printed parts and components sourced from online stores, including Adafruit and SparkFun. Since everything would eventually be hidden inside K-9's body, Fitz did all his prototyping on a separate bench using

> ❚❚ **K-9 is a combination of 3D-printed parts and components sourced from online stores** ❚❚

a different Raspberry Pi, before transferring everything to the ready-assembled setup inside the robot. This requires some duplication of hardware, but makes life easier for a project with plenty of iterative revisions and "continuous modification and expansion." Utilising the various GPIO connections, to both read inputs from

◀ Large bright buttons give K-9 an authentic 1970s look

▼ K-9's scarf shows his allegiance to his favourite Doctor, clips of whom might well appear on-screen imminently

Give K-9 new tricks

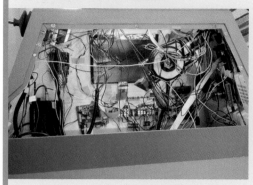

01 Download templates for K-9's body and an electronics schematic from **magpi.cc/k9bodyconstruction** to either 3D-print or laser-cut. You will need at least a Raspberry Pi 2, plus plenty of servos, and a fair amount of mechatronics knowledge.

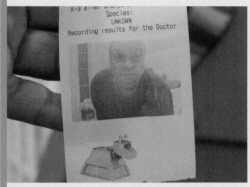

02 If you want to add the AI camera feature, you can adapt code from this photo booth project: **magpi.cc/photoboothdiy**. You'll need a thermal printer to output K-9's 'human or alien' verdicts.

03 Fitz used Python on his Raspberry Pi 3 to control the camera/printer and eye lights in K-9's head, and to play video clips on K-9's side monitor.

an external trigger and to run various external devices, was a big challenge. He estimates the build took three months of fairly focused work, with some light additions over the course of the following year. After an outlay of roughly $1000, the only real running costs are for the thermal paper on which K-9 prints photos of whoever is in front of him, along with a wry take on whether they might be human or alien.

You can see Fitz's K-9 robot in action at **magpi.cc/k9chronicles**. Expect a new post soon, as he's currently working on adding voice-recognition and Pico elements. 🎁

Maker Guides

108

124

120

154

154 Pico Iron Man Arc Reactor

Create a stunning Pico-based cosplay prop with this amazing arc reactor build

160 Retro gaming with Raspberry Pi Pico

You can play and make amazing games with Raspberry Pi Pico and RP2040

136

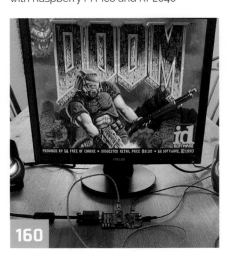

160

SERIOUS FUN

WITH ELECTRONICS

Discover the sheer joy of circuits, components, and physical computing. By **Lucy Hattersley**

With any luck, you've managed to get a wonderful Raspberry Pi computer. So, the question now is what to make with it?

Raspberry Pi isn't like other computers. In the computers of yore, all the components were highly visible and this made computing easier to understand.

Modern devices are made from glass and glue with all the interesting innards kept away from prying hands behind security screws and 'warranty void' warnings. Good for keeping little hands out; lousy for learning.

On every Raspberry Pi sit GPIO (general purpose input/output) connections. These pins enable Raspberry Pi to connect physically with electronics. Here is the real joy of Raspberry Pi. There are thousands of components you can use with your computer, from buttons and buzzers to small screens and sensors. You can recreate just about any gadget you own and bring your own ideas to life.

This feature is for those who have a Raspberry Pi and want to have some serious fun with electronics. We'll show you how to hook up wires, connect HATs, and get started on a wonderful electrical journey.

Jumper leads are used to connect Raspberry Pi GPIO pins to components on the breadboard (and to connect components to each other)

This white box is a 'breadboard', used to prototype simple circuits. Components are inserted into the holes (which connects them to nearby components)

GPIO pins are used to get input and send output to and from components

Components come in all shapes and sizes and perform a wide variety of tasks. Common components include buttons, LEDs, resistors, and buttons

Missing pins?

Raspberry Pi Zero models come with 40 GPIO holes, but Raspberry Pi Zero isn't populated with a 'header' (the physical pins that stick out). Some users solder a header directly to the GPIO (**magpi.cc/header**). If you are uncomfortable with soldering, you can use a Hammer Header and tap the pins (gently) into Raspberry Pi (**magpi.cc/hammerheader**).

KIT AND COMPONENTS

The kit you can use to learn electronics

BREADBOARD ↓

Experienced makers can get away with soldering components together with wires, but it's a faff and it's much faster to prototype circuits using a handy piece of plastic called a 'breadboard'.

Unlike its food namesake, an electronics breadboard is a plastic slab with a bunch of holes in it. At first glance, it looks pretty unfathomable, but it soon becomes easy to understand. See our How To Use a Breadboard tutorial (**magpi.cc/ breadboard**). In case you're wondering about the name, the first breadboards were wooden boards with rows of nails instead of holes.

TIP!

Remember: holes are connected in columns, aside from the split in the middle, so a component lead in A1 is electrically connected to anything you add to B1, C1, D1, and E1.

TIP!

Never try to cram more than one component lead or jumper wire into a single hole on the breadboard.

Terminal strips

On the breadboard are columns of (normally five) holes, called terminal strips. These are spaced 2.54 mm apart and underneath each column of five holes is a metal strip connecting that column. Components placed into a column are connected to each other as if they were physically wired together.

Power rails

Larger breadboards have strips of holes down the sides, typically marked with red and black or red and blue stripes. These are called 'power rails' and all of the holes in each rail are connected. These are used to provide common ground and power for a project. By connecting one ground hole to a ground pin on Raspberry Pi, all holes in the rail will act as ground. You can do something similar using a power pin if a circuit needs 3.3 V or 5 V power.

DIP spacing

In the centre of a breadboard, between the two columns of terminal strips, is a gap. This is usually the exact size to place a DIP (dual in-line package) chip. These IC chips can straddle the central division with a row of pins falling into holes on either side. Four-pin buttons are also the right size to straddle the gap.

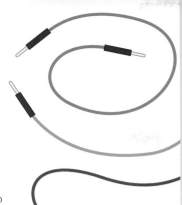

JUMPERS →

Jumper wires, also known as 'jumper leads' or 'jumper cables', connect components to each other, and Raspberry Pi's GPIO pins to the breadboard. Different types are used to connect to holes and pins.

- Male-to-female (M2F), which you'll need to connect a breadboard to the GPIO pins;
- Female-to-female (F2F), which can be used to connect individual components together if you're not using a breadboard; and
- Male-to-male (M2M), which is used to make connections from one part of a breadboard to another.

You'll need all three for more complicated projects. And you can pick up a pack like from most Raspberry Pi resellers (like The Pi Hut's Jumper Bumper pack, **magpi.cc/jumperbumper**).

RESISTORS →

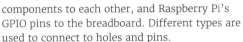

Resistors control the flow of electrical current and are available in different values measured using a unit called ohms (Ω). The more ohms, the more resistance is provided. For Raspberry Pi physical computing projects, their most common use is to protect LEDs from drawing too much current and damaging themselves or your Raspberry Pi; for this, you'll want resistors rated at around 330 Ω, though many electrical suppliers sell handy packs containing a number of different commonly used values to give you more flexibility. The more powerful the resistor, the dimmer the LED – be careful not to use one too strong or the LED light might not be visible.

BUTTONS →

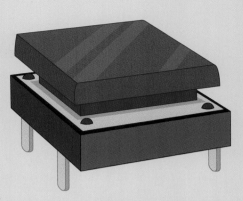

A push-button switch is used as an input device. You tell a program to watch out for it being pushed and then perform a task (button is pushed; turn on light, for example). They are commonly available with two or four legs – either type will work with Raspberry Pi.

It's much faster to prototype circuits using a handy piece of plastic called a breadboard

LEDS →

A light-emitting diode (LED) is an output device, a small light for your circuit that can be turned on and off via code. LED lights are found in many electronic gadgets, such as the light on a washing machine to let you know it's turned on. LEDs are available in a wide range of shapes, colours, and sizes, but not all are suitable for use with Raspberry Pi: avoid any which say they are designed for 5V or 12V power supplies. Instead pick lower voltages such as 1.2V, 3.8V, or the 2V LEDs found in this pack: **magpi.cc/ledpack**. The 'diode' part of LED means it can only be used one way around, so be sure to check the direction when using LEDs in your circuit.

Reading resistor colour codes

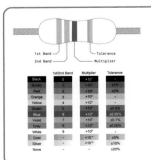

Resistors come in a wide range of values, from zero-resistance versions, which are effectively just pieces of wire, to high-resistance versions the size of your leg. Very few of these resistors have their values printed on them in numbers, though: instead, they use a special code printed as coloured stripes or bands around the body of the resistor. RS Components has a great guide to reading resistors (**magpi.cc/resistorsguide**).

BUZZER ↑

As you'd expect, a buzzer produces a buzzing noise. Inside is a pair of metal plates that vibrate against each other to make the sound. There are two types of buzzers: active and passive. Be sure to get an active buzzer, such as this one from Pi Hut, **magpi.cc/buzzer5v**, as these are the simplest to use.

UNDERSTANDING WIRING

Start putting together your components and learn to code

Electronic circuits are daunting for complete beginners. Fortunately, help is widely available, and it soon becomes a lot simpler than you imagine.

Once you have the breadboard and components, you should start wiring them up by following a simple tutorial (like the one at the end of this feature).

There's absolutely no shortage of electronic projects out there, and most of them guide you through attaching components to your breadboard, then using software such as the GPIO Zero Python library (**magpi.cc/gpiozero**).

There are a range of easy-to-follow electronics tutorials on the Raspberry Pi Foundation's website (**magpi.cc/electroniccomponents**).

Most tutorials include a wiring diagram. See the wiring diagram below. This provides a visual guide to how circuit is built. You'll find wiring diagrams like this throughout *The MagPi*, and similar ones used by the Raspberry Pi Foundation and other resources. If you're interested in creating your own, you use a program called Fritzing to make them (**fritzing.org**).

Components

The components are visually similar to real-life counterparts. Some, like the LED, visually demonstrate which way around they should go, so pay attention to the position of legs on components.

Simple Electronics with GPIO Zero

For more information on using GPIO Zero and learning to wire up circuits from wiring diagrams, take a look at our *Raspberry Pi Beginner's Guide*.
magpi.cc/BGbook

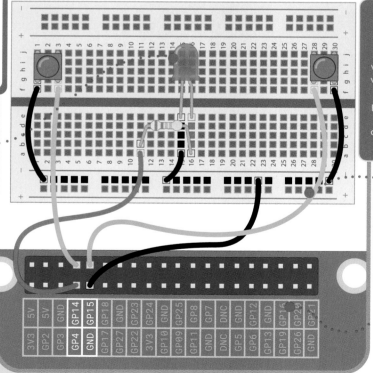

Wires

Jumper leads are represented by these coloured wires. The wires work the same, no matter which colour you use, but you'll often see red used for the power, and black connected to ground. The other colours are often used to represent various GPIO pin connections.

Pinout

Sometimes the diagram will have GPIO pin labels on the board, like this one. But these aren't printed on Raspberry Pi and many diagrams include a pinout separately. If you ever need a guide to the pins, you can find one here: **magpi.cc/pinout**.

PUT A HAT ON IT

Take the hassle out of circuit building with HATs

HATs (Hardware Attached on Top) are pre-built circuit boards with components and parts. They are designed to connect to the 40-pin GPIO header on Raspberry Pi, and are easy to set up and get started. Each HAT has an EEPROM on board with the software installation needed for the HAT to work, so all you need to do is plug it in and start using the electronics on board. Here are some fun HATs to try out. You'll also see pHATs, which are designed for Raspberry Pi Zero.

INKY WHAT ↑

Electronic ink displays, of the kind found in e-readers, are a great attachment to Raspberry Pi. Pimronoi's Inky wHAT (£52/$70) is a cut above with a large 400 × 300 three-colour screen (with a choice of red or yellow accent alongside the black and white). Take a look at our starter tutorial (**magpi.cc/inkyhello**) on using Inky wHAT.
magpi.cc/inkywhat

JAM HAT →

Designed to make starter electronics easier in the classroom, JAM HAT is packed with LEDs, buttons, and a buzzer. These components can be used with a range of beginner projects, like building a traffic light system with a button crossing. Students can focus on the code without spending the whole class wiring up the components. Of course, that's half the fun for us!
magpi.cc/jamhat

EXPLORER HAT PRO →

Put a breadboard on top of your Raspberry Pi with Pimoroni's Explorer HAT Pro (£20/$27). Alongside the mini breadboard are a range of inputs and outputs, including capacitive touchpads (that can be used as buttons), coloured LEDs, analogue inputs, and motor drivers. What's great about the Explorer HAT is that it's been around for years and is backed up with a custom Python code library and a bunch of examples on Pimoroni's learn page (**learn.pimoroni.com**).
magpi.cc/explorerhatpro

WEATHER HAT ↓

This is a neat solution for attaching climate and environmental sensors to Raspberry Pi. Wind/rain sensors are attached to the RJ11 connectors. Information is displayed on the 1.54in LCD screen. You can get a whole kit, including HAT and a wind vane, anemometer (wind speed), and rain gauge. Pimoroni has a guide for getting started (**magpi.cc/weatherhatstarter**).
magpi.cc/weatherhat

SENSEHAT →

Designed for Raspberry Pi's space program, Astro Pi, Sense HAT allows Raspberry Pi to sense the world around it. Two have been on-board the ISS since 2017. And, every year, students around the world use them to perform experiments in space. Down to earth, the Sense HAT has orientation, pressure, humidity, and temperature sensors, along with an LED Matrix and joystick control. Best of all, it is backed by hundreds of documented experiments and our Sense Hat Experiments book (**magpi.cc/sensehatbook**), with projects such as a Gravity Simulator, Magic 8 Ball, Pixel Pet, and Data Logger.
magpi.cc/sensehat

There's absolutely no shortage of electronic projects out there

ZEROSEG →

One of the more popular HAT projects combines two four-digit LED unit displays with two buttons. The result is a small HAT that can be used to create scrolling tickers, such as news displays, stock market results, and short moving messages.
magpi.cc/zeroseg

GET STARTED

While you can buy components separately, it is often easier to pick up a starter kit

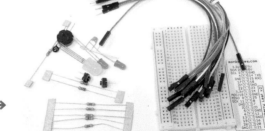

MONK MAKES →

Simon Monk has been putting together starter kits and tutorial guides since 2013, and his Raspberry Pi collection is one of the finest. You'll find a tutorial by Simon overleaf based upon his Project Box for Raspberry Pi (£12/$9). Inside you'll find a breadboard, jumper wires, LEDs, resistors, push-buttons, buzzer, and both a thermistor and phototransistor. There's also a booklet with ten products. We also like his 'Leaf', a plastic guide to GPIO pins that fits over Raspberry Pi and makes it easier to identify which wire goes where.
monkmakes.com

The CamJam EduKit is a stalwart classic of Raspberry Pi electronics

RASPBERRY PI 4 ULTIMATE KIT ↓

If you are starting from scratch, and want everything you need to be delivered in one place, then CanaKit's Ultimate Kit (£95/$129) is a great option. Inside the box are a Raspberry Pi 4 Model B, power supply, case, microSD card, cables, and a breadboard with electronics parts.
magpi.cc/canakitultimate

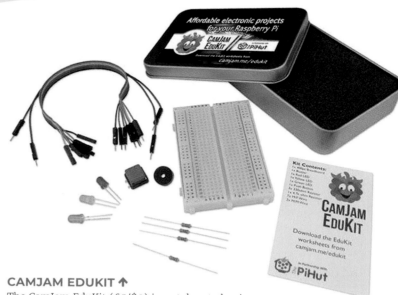

CAMJAM EDUKIT ↑

The CamJam EduKit (£5/$9) is a stalwart classic of Raspberry Pi electronics, and many a maker cut their teeth here. Inside the tin is a breadboard, resistors, LEDs, button, buzzer, and jumper leads. Good parts, and good value. 🅜
magpi.cc/edukit

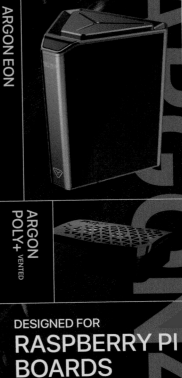

Get started with
electronics with
Raspberry Pi

Raspberry Pi is a great way to start learning about electronics.
Here's how to take your first steps in this fascinating subject

WRITER

Simon Monk

Simon divides his time between writing and designing products for MonkMakes Ltd. His books include *Programming Raspberry Pi* (TAB) and *The Raspberry Pi Cookbook* (O'Reilly). He has sold over 700,000 books in ten different languages.

@simonmonk2

You'll Need

▶ Raspberry Pi with GPIO header

▶ Breadboard

▶ Male-to-female jumper leads, LED, button
magpi.cc/ projectbox1

One of the great things about Raspberry Pi, is the inclusion of the double row of pins called the GPIO connector.** This enables you to connect external electronics to your Raspberry Pi and use code to control things like LEDs and buzzers, as well as reading values from sensors or detecting when switches are pressed.

In this tutorial, you will use the popular MonkMakes Project Box 1 for Raspberry Pi (**monkmakes.com**) to get started with electronics. This kit contains the LED, jumper leads, breadboard, and buttons we will use to explore electronics. You can pick up these components separately, and other kits are available.

The Leaf template makes it easier to identify GPIO pins

No soldering will be required, as you will be using a solderless breadboard to make your electronic circuits, and then connect them to your Raspberry Pi using jumper wires.

01 **LED and resistor on breadboard**
Place an LED, resistor, and switch onto the breadboard as shown in **Figure 1** (overleaf). You can identify which resistor to use by its stripes (RS Components has a good guide to resistors, **magpi.cc/resistors**). Use one of the 470 Ω resistors that have yellow, purple, and brown stripes.

One of the leads of the LED is slightly longer than the other. This is the positive lead and should be on row 3 of the breadboard, as shown. It does not matter which way around the resistor or switch go, but make sure that the switch pins run top to bottom.

02 **Fit the Leaf**
Place the Raspberry Leaf GPIO template from the kit over the GPIO header pins, so that the text on the Leaf that says 'Raspberry Leaf' is to the outside of Raspberry Pi. This allows you to easily identify which GPIO pin is which.

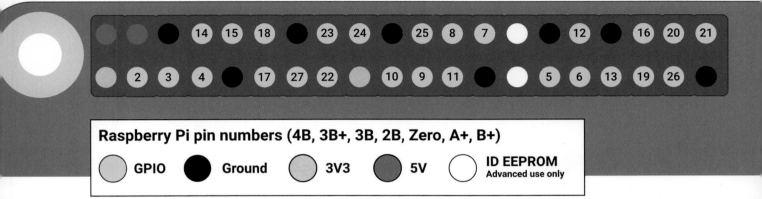

Raspberry Pi pin numbers (4B, 3B+, 3B, 2B, Zero, A+, B+)

⬤ GPIO ● Ground ⬤ 3V3 ⬤ 5V ◯ ID EEPROM Advanced use only

03 Connecting to Raspberry Pi

Use two female-to-male jumper leads (**magpi.cc/mfjumpers**) to make the connections from your Raspberry Pi to the breadboard. It does not matter what colour leads you use, but a convention is to use red for plus volts and black or blue for 0V (GND/ground).

Connect from 5V on Raspberry Pi to row 2 on the left side of the breadboard; from GND on Raspberry Pi's pins to row 5 on the right side of the breadboard, which will connect to the negative side of the LED.

There are eight pins on the Raspberry Pi's GPIO connector labelled GND and it does not matter which you use. Similarly, there are two 5V pins and either can be used.

04 Try it!

This simple circuit only uses Raspberry Pi to provide power to the circuit. The 5V supply from Raspberry Pi connects to one end of the push switch. If the switch is pressed, then electrical current can continue flowing and will flow first through the resistor and then through the LED before returning back to Raspberry Pi's GND connection. When the button is pressed and the current flows, the LED will light.

The resistor has the job of restricting the flow of current, as otherwise too much current would flow through the LED and it would burn out.

05 Two switches

As it stands, we are not getting anything from Raspberry Pi that we couldn't from a battery.

So, let's include Raspberry Pi in the action so that it can monitor two switches and when they are pressed, use them to alter the brightness of the LED.

Pull all the jumper leads out from Raspberry Pi and breadboard, being careful that none of the metal ends of the jumper wires touch.

Remake the breadboard so that there are now two switches as well as the LED and resistor connected to the breadboard. This will involve moving the first switch and adding a new one.

06 Raspberry Pi in control

Use six female-to-male jumper wires, as shown in **Figure 2**, to connect the LED (and its resistor) and switches. Notice that now the LED and two switches are each connected separately to Raspberry Pi GPIO pins and GND pins. The LED is

▲ Use this pinout guide to identify the GPIO pins. The yellow circles pins are used in code using the corresponding BCM number, black pins are for ground, red pins provide constant 5V power, while the orange pin provides 3V power. The white pins are reserved for add-on hardware

▼ You can edit the program and run it using Thonny

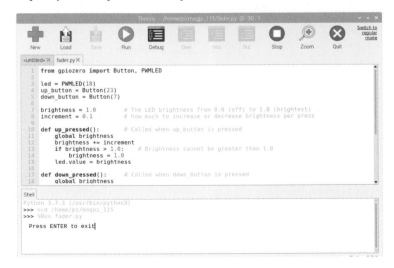

Figure 1

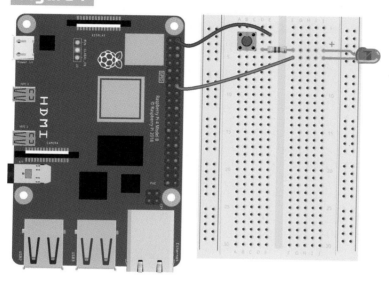

▲ **Figure 1** Here we use Raspberry Pi like a battery, and supply 5 V to provide power to an LED (via a resistor to protect it)

Figure 2

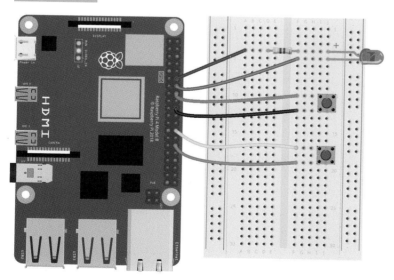

▲ **Figure 2** Wiring the LED fader project. Make sure you get the LED the right way round and the wires in the right place

connected to GPIO 18, which will act as an output that we can use to control the LED's brightness. The two switches are connected to GPIO pins 23 and 7. These GPIO pins will be acting as inputs, so that our code can detect when the switches are pressed and carry out some action.

07 Get the code!

As it stands, nothing will happen when we press the buttons because we need a program to be running on our Raspberry Pi that can monitor the buttons and do things when they are pressed. To fetch that code from GitHub, first open a Terminal window. This can be found in the Accessories section of Raspberry Pi OS's start menu.
Make sure you have a connection to the internet and run the command:

```
git clone https://github.com/simonmonk/
magpi_115.git
```

This will fetch the program into a directory called **magpi_115**. You will also find the **fader.py** code listing in this article.

08 Run the code

To run the code, first change directory and then run the program by typing the commands below into the Terminal:

```
cd /home/pi/magpi_115
python3 fader.py
```

If everything is OK, you should see the message 'Press ENTER to exit'. If you get error messages, go back to Step 7 and make sure the code downloaded OK.

09 Try it out!

When the program first runs, the LED should be at maximum brightness. Press the lower of the two buttons on the breadboard and you

fader.py

> Language: **Python**

```python
001.  from gpiozero import Button, PWMLED
002.
003.  led = PWMLED(18)           # variable brightness LED using pin 18
004.  up_button = Button(23)
005.  down_button = Button(7)
006.
007.  brightness = 1.0           # The LED brightness from 0.0 (off) to 1.0 (brightest)
008.  increment = 0.1            # how much to increase or decrease brightness per press
009.
010.  def up_pressed():          # Called when up_button is pressed
011.      global brightness
012.      brightness += increment
013.      if brightness > 1.0:   # Brightness cannot be greater than 1.0
014.          brightness = 1.0
015.      led.value = brightness
016.
017.  def down_pressed():        # Called when down_button is pressed
018.      global brightness
019.      brightness -= increment
020.      if brightness < 0.0:   # Brightness cannot be less than 0.0
021.          brightness = 0.0
022.      led.value = brightness
023.
024.  up_button.when_pressed = up_pressed # link up_pressed to up_button
025.  down_button.when_pressed = down_pressed
026.
027.  led.value = brightness              # so that the LED is lit even if buttons not pressed
028.
029.  input("Press ENTER to exit")    # avoid the program finishing as soon as it starts
030.
```

should see the LED start to dim. Press it some more times and the LED will go off altogether. Pressing the other button will increase the brightness.

When you are ready to quit the program, just press **ENTER** on your keyboard.

10 Looking at the code

If you want to inspect or change the code for this project, run Thonny from Raspberry Pi OS's menu (in the Programming section) and click the Load button, navigate to **fader.py**,

and open it. You can also run the program from Thonny by clicking on the Run button.

Try changing the value of the variable `increment` from 0.1 to 0.2. You will notice that the brightness changes in larger increments.

11 What next?

We have used a few components from Project Box 1 for Raspberry Pi. This kit contains the parts and instructions for lots of other interesting projects to get you started with coding and electronics.

Top Tips 👍

Unplug Raspberry Pi

It's easy to accidentally connect wires that shouldn't be connected and this could damage your Raspberry Pi (although it's unlikely). So, it's a good idea to unplug your Raspberry Pi when moving components about.

LED polarity

If the LED doesn't light when it should, then check that it is the right way around.

Get started with
the command line

Learn how to use text-based commands in Raspberry Pi OS

Phil King

Long-time contributor to *The MagPi*, Phil is a freelance writer and editor with a focus on technology.

@philkingeditor

Get Started with the Command Line book

To discover more useful commands and tips for using the command line on Raspberry Pi, check out our Conquer the Command Line book, available as a free PDF:
magpi.cc/commandline

The standard version of Raspberry Pi OS has a user-friendly graphical user interface, but there are a few limitations to what you can do with it. That's where the command line comes in, enabling you to get under the hood of the system and unleash the power of text-based commands.

You can access the command line in several ways. The simplest is to open the Terminal app in Raspberry Pi OS (**Figure 1**) – either by clicking its icon (a black rectangle) in the taskbar or opening it from the menu – it's in the Accessories category. Alternatively, you can open a virtual console by pressing **ALT**+**CTRL**+**F1**. You'll also be taken straight to the command line if you install the Lite version of Raspberry Pi OS, without the desktop – or if you opt to 'Boot to CLI' in the Raspberry Pi Configuration tool.

Another command way to access the command line is to use SSH (Secure Shell) – see the guide at **magpi.cc/ssh** – to access your Raspberry Pi remotely from another computer, which is often very useful.

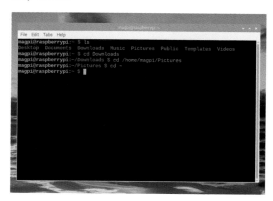

▲ **Figure 1**: You can access the command line from the Terminal app, as shown here, or by switching to a virtual console or SSHing in from another computer

01 First steps

When you first access the command line, via the Terminal or other means, you will see a prompt starting with your username followed by '@' and the hostname (raspberrypi by default) with a colon. For instance:

```
magpi@raspberrypi:
```

After this, you'll see the current directory you're in, followed by a '$'. You'll start off in the user's home directory, indicated by '~'.

'Directory' is another word for folder and they correspond to the ones seen in the File Manager on the desktop. So, within your user's home directory (which is located at **/home/[username]**), you'll find the usual default directories such as **Downloads** and **Pictures**. To see what's in the current directory, we use the `ls` command:

```
ls
```

This is short for 'list' and shows the contents of the current directory.

02 Navigating directories

To change to a different directory, the `cd` command is used. For instance:

```
cd Downloads
```

You will now see '~/Downloads' before the '$' in the prompt. Note that to visit another directory in your home folder from here, you'll have to enter the full path, such as:

```
cd home/magpi/Pictures
```

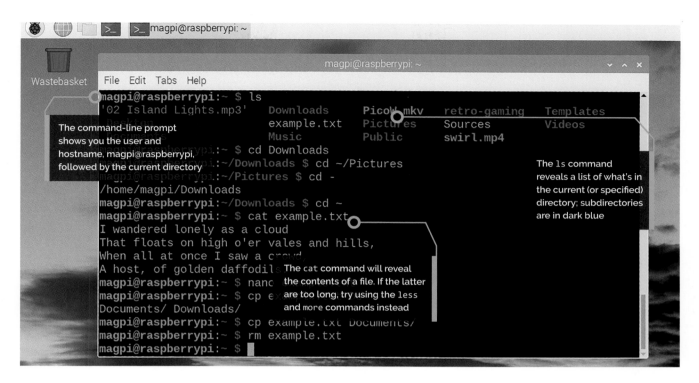

The command-line prompt shows you the user and hostname, magpi@raspberrypi, followed by the current directory

The `ls` command reveals a list of what's in the current (or specified) directory; subdirectories are in dark blue

The `cat` command will reveal the contents of a file. If the latter are too long, try using the `less` and `more` commands instead

Or you can use the shorthand of '~' in place of 'home/magpi':

```
cd ~/Pictures
```

In addition, `cd ..` takes you up one directory level (to the parent directory) while `cd -` returns you to the previous directory.

If you forget what directory you are in, you can find it with:

```
pwd
```

This stands for 'print working directory'.

03 File manipulation

Now we can find our way around the file system, let's create a new text file using the `touch` command. In the home directory, enter:

```
touch example.txt
```

Enter the `ls` command and it will show up in the list. To edit it, we can use the nano text editor:

```
nano example.txt
```

Enter a few lines of text (**Figure 2**) and then close it with **CTRL+X** followed by **Y** and **ENTER** to save the changes. We can see what's in a text-based file using the `cat` command.

```
cat example.txt
```

If we want to move or copy the file to another directory, we use the `mv` or `cp` command respectively, followed by the path of the target directory. For instance:

```
cp example.txt ~/Documents
```

If you `cd` to the **Documents** directory and `ls`, you will see the copy of the file there. You can delete a file using the `rm` command:

```
rm example.txt
```

To search for a file, you can use the `find` command with a selected directory, `-name` option, and filename (which can include the * wildcard for 'any characters'):

```
find /home/magpi -name "e*.txt"
```

Top Tip 👍

Auto-complete

The **TAB** key is your friend on the command line. Start typing a command, file, or directory name and then press **TAB** to either auto-complete it or list the possibilities. You can also use the up and down arrows to explore your command history and quickly re-enter previous commands.

◀ **Figure 2**: The nano text editor enables you edit text-based files, including programs that you can run

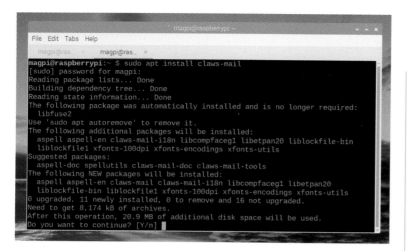

Figure 3: You can install a wide variety of software from the command line using the APT package manager

Top Tip

Case-sensitive

Note that file and directory names are case-sensitive, so entering `cd downloads` won't take you into the **Downloads** directory, and will elicit an error unless there is one called 'downloads'.

Figure 4: You can run Python programs from the command line using the `python` or `python3` command

04 Making directories

You can also create a new directory with the `mkdir` command. For instance:

```
mkdir Test
```

You can delete a directory with the `rm` command, so long as you add the `-R` switch (short for recursive) to delete everything in it:

```
rm -R Test
```

Be careful with `rm` command as deleted files are not recoverable!

05 Installing software

While there is now an updater tool that appears in the taskbar of Raspberry Pi OS desktop when software updates are available, you can also update your software packages from the command line. To download the latest information on the configured software packages, enter:

```
sudo apt update
```

Here, we are using `apt` for the APT package manager used in Raspberry Pi OS with its `update` command. Note that we need to prefix it with `sudo`, to gain the required superuser privileges – you will be asked to enter your password.

This command is often used before upgrading all your software packages:

```
sudo apt upgrade
```

Or when installing new software. To install an example software package (that is in the APT repository), we use the `sudo apt install` command with the package name. For instance, to install the Claws Mail email client:

```
sudo apt install claws-mail
```

You will be shown information for it and asked if you want to proceed – enter **Y** to confirm (**Figure 3**). The installation process will then begin. Once it's installed, you will find Claws Mail in the desktop applications menu, under Internet. Alternatively, you can run it from the command line (although it will open in GUI window):

```
claws-mail
```

To remove a software package, you can use `sudo apt remove` followed by its name.

06 Download and install

Some software may not be available from the APT repository, in which case you will need to download it (making sure it works with the ARM architecture for your Raspberry Pi) and install it. Instructions will typically be given on the website for the software. This may involve using a `curl` command to download it and run a script to install it, or a `wget` command with the URL for the file (typically a zip). You may also be asked to clone the GitHub repository for a project, such as:

```
git clone https://github.com/
themagpimag/retro-gaming.git
```

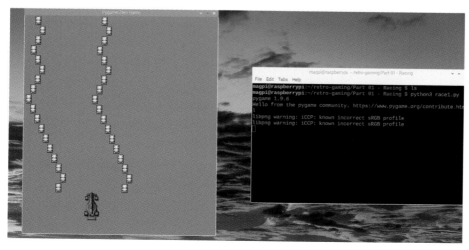

Once it has downloaded, you'll find the **retro-gaming** directory containing subdirectories with .py Python files that you can run. For instance:

```
cd /retro-gaming/'Part 01 - Racing'/
python3 race1.py
```

See **Figure 4**. To stop a program running, you can press **CTRL+C**.

07 Command-line internet

You can even perform web searches directly from the command line. First, install the surfraw tool:

```
sudo apt install surfraw
```

Surfraw works with helpers called 'elvi'. To see a list of them, enter:

```
surfraw -elvi
```

Try searching for something with one, using the `sr` command.

```
sr duckduckgo Raspberry Pi
```

The results will open in a window for the default web browser, Chromium. You can, however, also view many websites in the command line itself, using the w3m text-based web browser. Install it with:

```
sudo apt install w3m
```

▲ **Figure 6**: Playing a video in ASCII mode – can you tell what it is yet?

Then use it to view a website:

```
w3m raspberrypi.com
```

For a weather forecast, try:

```
w3m wttr.in/Poole
```

You'll see the forecast for different days, complete with ASCII graphics (**Figure 5**). Other text-friendly sites include Google.com and Wikipedia.org.

08 Audio and video

It's possible to play audio and even video from the command line using the mpv media player. Download or copy across an MP3 file and then play it with:

```
mpv example.mp3
```

You can `sudo apt install` the youtube-dl tool and then use it download a video from a link, or just use the 'v=' code at the end of the URL for it, such as:

```
youtube-dl QIdyTlmdVW8
```

This displays a Pico W video. Note that it can take quite a while to download via the proxy, however. Like any other video file, you can then use the `mpv` command with the `-vo=caca` option to play it in super-low-res ASCII (**Figure 6**)! Press **Q** to quit playing the video. 🄼

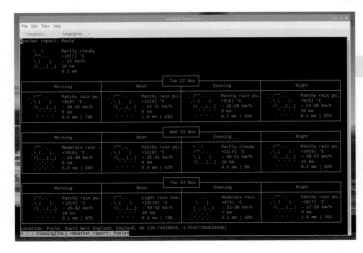

◀ **Figure 5**: It's possible to browse the web from the command line. Some sites, such as wttr.in, are designed for text-based browsers

Top Tip

Software search

To search for software packages in the APT repository, use `apt-cache search` followed by the search term. Alternatively, download the Sources.gz file from **magpi.cc/ aptsourcesbullseye** and use gunzip to extract it for the full list of software packages.

Raspberry Pi high-end audio

It's time to up your audio game on our favourite computer!
PJ Evans takes us through the options. This might get loud

From the very beginning of Raspberry Pi, **audio has been one of its most popular applications**. Raspberry Pi's small form factor and fixed parts lend themselves perfectly to sitting in a living room or kitchen. Although the Raspberry Pi Model A and B computers feature built-in audio, it wasn't long before more advanced HATs (Hardware Attached on Top) appeared, raising the output quality to something that would make any self-respecting audiophile drool. Add in amazing audio management software, and we had a rival for top-end home audio systems.

In this special feature, we're going to look at the audio hardware available and how to get the best out of it.

USB Audio Adaptor

The Pi Hut ▶ £4.50

The logical step up from Raspberry Pi's built-in audio is to add a USB audio adaptor. This is also the neatest low-cost solution for adding audio to Raspberry Pi Zero (with a USB micro-to-A adaptor, **magpi.cc/microusbadaptor**) and a simple method for multi-channel projects, as you can add as many adaptors as you like. Many different versions are available so be sure to check that the one you are considering will work with your choice of operating system. Most (including the one linked to here) will work without any configuration or driver installations. This will also add the ability to record audio on your Raspberry Pi.
magpi.cc/usbaudio

Your audio options

Built-in audio

Raspberry Pi ▶ Free!

When choosing your audio solution, don't forget that all Raspberry Pi Model A and Model B variants come with audio baked right in. All feature a 3.5mm socket that provides stereo audio at line-out levels. On later models, such as Raspberry Pi 3 Model B and Raspberry Pi 4 Model B, this is a four-pole connector (TRRS; Tip, Ring, Ring, Sleeve) that also provides composite video. So, if you're not after high-quality audio and just want to make some noise, just add an amplifier and you're set. Raspberry Pi Zero computers do not feature 3.5mm audio out, so only its larger cousins will do. Suitable cables are widely and cheaply available.
magpi.cc/raspberrypi4

▲ Need basic stereo audio? The Raspberry Pi 4B has you covered with a standard 3.5mm headphone socket

◄ Simple but effective, a USB audio adaptor adds sound and recording capability to any Raspberry Pi

▲ Pimoroni's range of audio HATs cover all common use-cases

Pirate Audio
Headphone Amp

Pimoroni ▶ £20

With a name like Pirate Audio, it can only be our friends from Sheffield-on-Sea, Pimoroni. Its Pirate Audio range is similar to the HiFiBerry DAC but with more features. We've chosen the headphone version here, but line-out, 3W amp and even built-in speaker versions are available. All four Raspberry Pi Zero-sized HATs feature a 1.3-inch IPS screen and four control buttons, making them perfect for on-the-move audio. A comprehensive online guide takes you through installation including a full audio solution based on Mopidy (**magpi.cc/mopidy**), so you can get running right away.
magpi.cc/pirateaudioheadphone

PecanPi

Orchard Audio ▶ $350

Orchard Audio has firmly established itself at the top of the pile when it comes to Raspberry Pi audiophiles. A seemingly relentless dedication to sourcing the best components and cutting absolutely no corners when it comes to the design of their DAC boards has resulted in the kind of performance no one could ever have associated with the humble Raspberry Pi. The DIY version of its signature PecanPi is a fully-loaded HAT suitable for studio use with twin DACs and XLR outputs. Yes, the price is eye-watering, but you get what you're paying for. If only the best is good enough, you've found your product.
magpi.cc/pecanpi

HiFiBerry DAC+ Zero

HiFiBerry ▶ £18

Computers don't understand audio as we hear it. Instead, they use a Digital-Analogue Converter (DAC) to turn a digital signal into something we can hear. The quality of the DAC integrated circuit is the single most important factor in producing great audio quality. Adding a dedicated DAC to your Raspberry Pi is the best bang-for-buck upgrade you can get.

At just £18, this HAT produces great line-out audio quality and is perfect for Raspberry Pi Zero projects. Several variations are available that add digital audio and even small amplifiers. A great range of cases means it is perfect for home audio projects.
magpi.cc/hifiberrydaczero

▲ A low-cost but high-quality DAC can add greats sound reproduction to a Raspberry Pi Zero

▼ Perfect for your voice assistant project, the Respeaker's far-field microphones capture great quality speech for more reliable recognition

Respeaker v2

Seeedstudio ▶ £25

So far we've mostly covered audio output devices, but what about input? The easiest solution is to get a USB audio adaptor, but you'll also need a suitable microphone and amplifier if you want to record voice. A popular use of Raspberry Pi in the home is to build a voice assistant. If you fancy trying to build your own, this HAT-based audio recording device is everything you need. An array of four microphones with far-field capability can capture voice commands from 5m away. Resources and tutorials are available from ReSpeaker (respeaker.io) to help you build your own Alexa or Siri.
magpi.cc/respeaker

◀ Small but mighty. The PecanPi produces studio-grade sound

Top Tips 👍

Other audio types

There are plenty of other ways to make noise with a Raspberry Pi. A wide range of buzzers and sirens are available. Always be careful with volume!

Get what you need

Don't be tempted to splash out on the latest and greatest if all you want to do is make a buzz.

Set up a whole-house audio system

Whether it is a single room or everything including the loo, here's how to get audio in every room with Raspberry Pi

WRITER

PJ Evans

PJ is a writer, software engineer and a very bad audiophile. His prized Val Doonican collection can be listened to in pristine quality.

@mrpjevans

Just imagine being able to listen to your music anywhere in your home, in perfect sync as you move around. Such audio systems do exist, but can cause serious damage to your bank account. However, it is now possible to build just as good a system with nothing but Raspberry Pi computers and some incredible open-source software. Add a DAC HAT for a boost in sound quality, perhaps a cool case and speakers, and you're good to go. You can use streaming services such as Spotify too.

01 Install your DAC and configure

If you are using a DAC HAT, now is the time to install it. Follow the manufacturer's instructions closely and ensure your Raspberry Pi operating system is fully up-to-date. If you're using Raspberry Pi OS Lite, you will need to use `alsamixer` to enable the card (or at least turn the volume up). Once you're ready to test the

▲ Without its case you can see how the DAC+ Zero is assembled with Raspberry Pi Zero

audio and have connected the DAC to some active speakers, here's a simple command to check everything is working:

```
speaker-test -c 2
```

This will play white noise through the left and right channel. If you can hear it, you're good to proceed.

02 Install Mopidy

Now that we have sound, the next step is to install software to control and manage our library of music. Mopidy is an excellent choice for this and comes with Iris, a beautiful web interface. To install Mopidy follow the commands in **mopidy.txt** (press return after each line). This will install everything you need. Now, to access Mopidy remotely, edit the config file:

```
sudo nano /etc/mopidy/mopidy.conf
```

At the end of the file, add the code in listing **mopidyconfig.txt**. Now save it, then start the server:

```
sudo systemctl start mopidy
```

You should now get a response on **<ip address>:6680/iris/**

03 Load up some music

Your music folder is **/home/pi/Music**, and you can now transfer your music library to that

The HiFiBerry DAC+ Zero provide high-quality output through standard RCA connectors

Mopidy Iris provides a beautiful user interface to all your music

Warning!
Volume!

Always be careful when testing audio out, especially if amplified. Excessive volume can cause lasting hearing damage.

magpi.cc/hearingloss

" Add a DAC HAT for a boost in sound quality "

directory. It is common to use an Artist/Album/Tracks pattern. Mopidy will inspect the metadata in the files and catalogue accordingly. Once you're ready, ask Mopidy to scan the folder:

```
sudo mopidyctl local scan
sudo systemctl restart mopidy
```

If you can't immediately see your new files, try Browse > Local Media to locate them. You now have a fully featured, remote-controlled media player. Modipy has a host of plugins, so you can add services such as radio and Spotify.

04 Add Snapcast

Snapcast is an open-source multi-room streaming system that provides proper in-sync playback without loss of quality. To make it work, we re-route the Mopidy output stream to the Snapcast server, which then relays the signal to Snapcast clients, including locally. To install Snapcast server enter the following in Terminal:

```
wget https://github.com/badaix/
snapcast/releases/download/v0.26.0/
snapserver_0.26.0-1_armhf.deb
  sudo dpkg -i snapserver_0.26.0-1_armhf.deb
```

And the client…

```
wget https://github.com/badaix/
snapcast/releases/download/v0.26.0/
snapclient_0.26.0-1_armhf.deb
  sudo dpkg -i snapclient_0.26.0-1_armhf.deb
  sudo apt -f install
```

A dependency error on step 2 can be ignored.

Top Tips 👍

Groups

Snapcast supports 'groups', so you can have many different players playing back in sync. You can even have different music streams playing simultaneously.

▼ The Iris interface uses metadata from your music files to produce an interface that works equally well on desktop and mobile

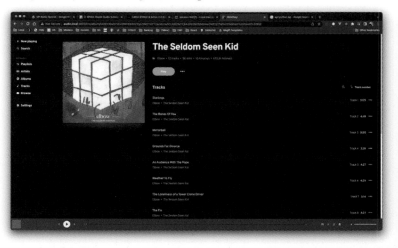

mopidy.txt

DOWNLOAD THE FULL CODE:

⬇ **magpi.cc/mopidytxt**

> Language: **BASH**

```
001.    wget -q -O - https://apt.mopidy.com/mopidy.gpg | sudo apt-key add
        -
002.    sudo wget -q -O /etc/apt/sources.list.d/mopidy.list https://apt.
        mopidy.com/bullseye.list
003.    sudo apt update
004.    sudo apt install mopidy python3-pip
005.    sudo adduser mopidy video
006.    sudo pip3 install Mopidy-Iris
007.    sudo pip3 install Mopidy-Local
008.    sudo sh -c 'echo "mopidy ALL=NOPASSWD: /usr/local/lib/python3.9/
        dist-packages/mopidy_iris/system.sh" >> /etc/sudoers'
009.    mkdir -p ~/Music
010.    sudo systemctl enable mopidy
```

mopidy_config.txt

DOWNLOAD THE FULL CODE:

⬇ **magpi.cc/mopidyconfigtxt**

> Language: **BASH**

```
001.    [http]
002.    hostname = 0.0.0.0
003.
004.    [audio]
005.    output = alsasink
006.
007.    [local]
008.    media_dir = /home/pi/Music
009.    sudo systemctl enable mopidy
```

05 Configure Snapcast

To enable Snapcast, go back to the web interface and select Settings. Click on the Snapcast icon, and click 'Enabled'. If it doesn't work immediately, change the 'Host' setting to the full name of your server. Once it says 'Connected', Snapcast is now broadcasting all music you play over your local network. Make sure playback works as normal, although it may take an extra second or two to start as it syncs. All that remains is to create additional 'nodes' on your whole-house audio system to accept the Snapcast stream.

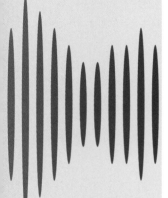

Top Tips 👍

Auto-scan your files

Mopidy won't detect new files automatically, so run `sudo mopidyctl local scan` regularly, or even as a nightly cron job.

06 Create additional music players

You can use any device that can run Snapcast to receive your audio. If you're using a Raspberry Pi Zero and a DAC+, configure them as per step one, then just install the snapclient package as before, but not the snapserver.
The final step is to tell the system where to get the music stream:

```
sudo nano /etc/default/snapclient
```

Add the following:

```
START_SNAPCLIENT=true
SNAPCLIENT_OPTS="--host 192.168.0.4"
```

Replace 192.160.0.4 with the actual IP address of your server. Then restart:

```
sudo systemctl restart snapclient
```

Go back to the Mopidy interface and you should see the new player as a 'Group' in Snapcast. Add as many of these as you want, and enjoy your music anywhere. 🅜

Audio Software for Raspberry Pi

Now that you have your hardware sorted out, how do you control it?
PJ Evans hits the wheels of steel (or the command line, in this case)

I
t really doesn't matter how good your
Raspberry Pi audio setup is if you can't
get any sound out of it. From media player
software to studio-grade editing suites, there's a
lot available for you to play with once you've got
your headphones on. In this section, we will take
a look at some of the popular audio tools available
for Raspberry Pi OS. We've already covered Mopidy
in the previous tutorial, so here's a guide to some
of the other audio software packages available both
for playback and composition.

Audacity

audacityteam.org

Audacity is a real stalwart of the open-source
community. Lovingly improved over many years,
from its humble beginnings, the program now
reached the level of a studio-ready editing tool.
Proper non-linear multitrack editing, direct
recording and so many filters it will make your
head spin. The plug-in architecture ensures that if
there's an effect you need, chances are somebody
else has already made it. Whether it's a quick
import and mix down to mono, or your 20-track
opus, this is the software package you need. It
really is astonishing that it is completely open
source with no catch.

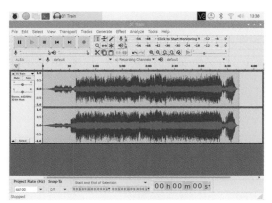

▲ A powerful user interface and multi-track editing

ffmpeg

ffmpeg.org

This command-line
application is truly the
swiss-army knife of audio
(and video) conversion. If all
you want to to is something
simple, such as convert a newly-captured WAV file
to MP3, there is no quicker way to do it:

```
ffmpeg -i my_kazoo_symphony.wav out.mp3
```

That's it. It will automatically detect quality and
channels and make sure everything sounds the
same. It supports a huge range of audio codecs,
including more complex multi-channel such
as AC-3 and DTS. An essential tool for anyone
who needs to process audio. Install with
`sudo apt install ffmpeg`.

▲ Not pretty to look
at, but behind the
command line are
powerful features

MuseScore

musescore.org

Something a bit different here. If you are a
composer or arranger, you may be interested in a
package that will help with sheet music notation.
Typically, commercial packages such as Sibelius
are the go-to tools, but one of the most popular
notation programs in the world is open-source.
MuseScore provides a wide
range of features including
MIDI-based input and
audio playback, as well
as being a fully featured
sheet music editor. It can
interact with several closed-
source packages as well. A
great asset for schools and
hobbyist composers.

▼ Sheet music notation
made easy with
this open-source
software package

Volumio

volumio.com

Billed as 'The Audiophile's Music Player', Volumio doesn't mess around. This is a dedicated, fully featured player squarely aimed at the high-end market. There is even a 'plug and play' OS version available right from Raspberry Pi Imager. With automatic support for a wide range of DACs (including many mentioned in this feature), Volumio will have you up and running in no time. Control is via a web interface and you can use a mobile version too, for relaxed sofa command. Its plug-in architecture allows for various music services such as Spotify or Tidal to work seamlessly.

▲ Volumio is a music player and library manager for the serious audiophile

Plex

plex.tv

Although focused on video, Plex still makes for a formidable audio player. You can run your own Plex server at home and have a single place for all your media. Plex will stream audio to any device accessing it via its web interface and many apps. Additionally, it can stream to any Google Chromecast or Sonos device with a PlexPass. Although not suitable for streaming to dedicated devices that do not support these protocols, Plex has a gorgeous interface and is friendly and intuitive to use. The ability to access your content remotely is a free feature (although some router config may be required).

▲ Access your music and videos anywhere with Plex's friendly interface

Top Tips 👍

Simple Command-Line

Raspberry Pi OS comes with the ALSA suite of tools for control, playback and recording from the command line

LibreELEC & OSMC

An honourable mention for these two optimised operating systems that make using Kodi, the popular media centre, quick and easy. Both available from Raspberry Pi Imager.

Sonic Pi

sonic-pi.net

No feature on Raspberry Pi audio would be complete without this amazing piece of software. Code? Music? Why not both? Live Coding is a discipline that uses code to produce sound, and therefore music. Supercollider, a popular engine amongst enthusiasts, is a complex and unwieldy beast, but Sam Aaron and the Sonic Pi team have tamed it in the Sonic Pi environment. For the beginner, the detailed and accessible tutorial is what makes Sonic Pi really stand out. Once mastered, it is capable of live performance and has often been used at events. A great and fun way to learn music theory.

▲ Sonic Pi means code that makes music. Detailed tutorials and a huge range of sounds

Great Audio Projects

Need some inspiration? Here are some of our favourite audio-based projects built on Raspberry Pi

Raspberry Pi Zero 2 Music Player

Drew Batchelor ▶ **magpi.cc/zero2mp**

When Drew needed a new music player for a kitchen, a creative side got the better of him and he decided to design an elegant 3D-printed case for the line-out version of Pirate Audio. It runs Volumio with a modified version of the Pirate Audio plugin to allow for better control operation and a clearer display.

RFID Vintage Boombox

Jorge Miar ▶ **magpi.cc/nfcboomboxyt**

Jorge's project takes the radio concept to the next level. As well as upcycling a fantastic 1980s boombox, an NFC reader has been added so you can control what is played back with cassettes. Place the tape of choice in the player and hit the play button.

Radio Globe

Jude Pullen ▶ **magpi.cc/radioglobe**

A beautiful concept that enables the user to explore over 2,000 radio stations from around the world based on the globe's location. This clever build plugs into Radio Garden (**radio.garden**) to provide the audio stream. Simply rotate the globe to anywhere to hear a nearby station live. Check out Jude's vlog of the build process (**magpi.cc/radioglobevlog**).

GTA Retro Radio Player

Raphaël Yancey ▶ **magpi.cc/gtaradio**

Featured before in these pages, the GTA radio player takes the radio stations featured in *GTA V* and allows the user to tune between them. The retro radio's innards are replaced with a Raspberry Pi and amplifier. A rotary encoder is read via GPIO and Python to move between stations.

Build a weather station
with a web dashboard

With the Pimoroni Weather HAT and Sensors Kit, you can upload your data to an easily accessible web dashboard

MAKER

Phil King

Long-time contributor to *The MagPi*, Phil is a freelance writer and editor with a focus on technology.

@philkingeditor

While it's possible to build a DIY Raspberry Pi weather station from separate components and sensors, Pimoroni's Weather HAT makes the process far simpler and easier. As well as on-board BME280 (temperature, pressure, humidity) and LTR-559 (light) sensors, the HAT features a Nuvoton microcontroller with a 12-bit ADC to read analogue signals reliably from external weather sensors connected via standard RJ11 ports. It even has a mini colour LCD screen to display readings.

The Weather Sensors Kit comprises three meteorological sensors: an anemometer to measure wind speed, a wind vane for direction, and 'tipping bucket' rainfall gauge. Alternatively, you may already have similar sensors or be able to source them elsewhere, but you'll need a couple of RJ11 connectors to plug them into the HAT – with wind speed and direction sensors routed through one connector. Either way, let's get started.

01 Set up Raspberry Pi

If you don't already have a recent version of Raspberry Pi OS written to your microSD card, use Raspberry Pi Imager (**magpi.cc/imager**) to do so from another computer. While you're at it, click the cog icon in Imager to access the Advanced Options. Here you can enable SSH (useful for remote operation later), set a username and password, and configure your WiFi connection. You may also want to change the hostname to something like 'weather.local', to make it easier to identify your weather station Raspberry Pi on the network (rather than using its IP address).

02 Mount the HAT

With your Raspberry Pi powered off, mount the Weather HAT on its GPIO header, with the body of the HAT over that of Raspberry Pi. You can use any Raspberry Pi model with a 40-pin header; we chose a Raspberry Pi Zero W for our

You'll Need

- Raspberry Pi
- Raspberry Pi OS
- Weather HAT + Weather Sensors Kit **magpi.cc/ weatherhat**

▲ Our weather station's cables go into the garage where a Weather HAT-equipped Raspberry Pi is located – an external weather-proof box would be better, however

setup and connected to it via SSH rather than using a monitor for setup. You can add standoffs and screws through the mounting holes next to the header (and the others if using a full-size Raspberry Pi) to secure the HAT more firmly.

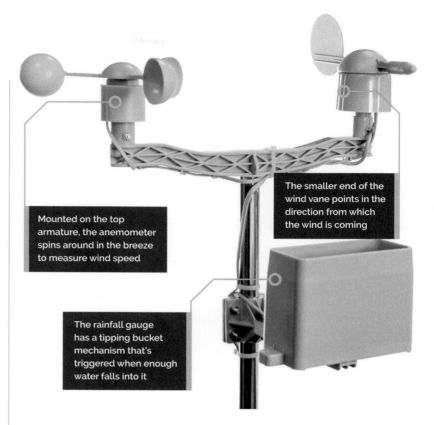

The smaller end of the wind vane points in the direction from which the wind is coming

Mounted on the top armature, the anemometer spins around in the breeze to measure wind speed

The rainfall gauge has a tipping bucket mechanism that's triggered when enough water falls into it

03 Install the software

With Raspberry Pi connected to a monitor and keyboard, open a Terminal window. Alternatively, access it remotely via SSH from another device. Enter the following commands to install the Weather HAT library:

```
git clone https://github.com/pimoroni/
weatherhat-python
cd weatherhat-python
sudo ./install.sh
```

This will automatically enable the I2C and SPI interfaces and install some additional software required for the HAT to work. In addition, the folder will include an **examples** subfolder of code examples. To be able to run these, you'll need to install some extra fonts and dependencies:

```
sudo pip3 install fonts font-manrope pyyaml
adafruit-io numpy
```

Note: If you're using the Lite version of Raspberry Pi OS, you may also need to install some additional software before the commands above will work. To do so, enter:

```
sudo apt install python3-pip git libatlas-
base-dev
```

▲ Mounted on a Raspberry Pi Zero (hidden beneath), the Weather HAT has on-board BME280 and light sensors

Now reboot Raspberry Pi with `sudo reboot`, for the changes to take effect.

04 Initial testing

With the software installed and Raspberry Pi rebooted, let's do a quick test of the HAT's on-board sensors. Change directory to the **examples** folder and run the main Python code demo:

```
cd weatherhat-python/examples
python weather.py
```

The default screen on the LCD will show several sensor readings. Since the external sensors are yet to be connected, you will only see those for the on-board BME280 and LTR-559 sensors for now: temperature, pressure, humidity, and light. We'll test the external sensors following assembly.

05 External sensors assembly

First, slide the two metal tubes together to make the mast. Then add the longer armature to the top of the mast, securing it with a screw. Mount the anemometer and wind vane securely on either side of the armature. The rain gauge fits onto a shorter armature further down the

Top Tip

Air & light

If using a weatherproof enclosure for Raspberry Pi, make sure it's well-ventilated from below. A transparent lid will also enable the light sensor to work.

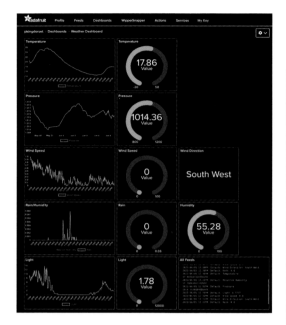

Our final Adafruit weather dashboard, using line charts, gauges, and a live stream for all feeds. You can design yours how you want it

Run the **weather.py** code example, as in Step 4. This time, you should see two extra readings at the top of the LCD's default screen: for rain (in mm/s) and wind (speed in m/s). Try spinning the anemometer and you'll see the wind reading increase. Similarly, try tilting the rainfall gauge up and down to make the internal bucket tip (you'll hear a clicking noise) and you should see the rain reading rise.

Pressing the X button on the HAT will change the numerical readings to graphs. Repeatedly pressing A will show specific displays for different sensors. With the wind display selected, test the wind vane by rotating it to different positions; the graph should change to show the direction.

mast, clear of the other sensors so they don't interfere with its operation. For more detailed build instructions, see SparkFun's assembly guide: **magpi.cc/weatherbuild**.

The wires from the wind sensors can be kept neat using the built-in cable clips on the top armature. The anemometer's RJ11 connector fits into a port on the underside of the wind vane. As you'd expect, the rain gauge connector goes into the 'Rain' port on the Weather HAT, while that coming from the wind vane goes into the 'Wind' port – note that this cable is thicker, as it carries four wires (for both wind sensors) instead of two.

07 Taking it outside

There's little point getting weather readings from inside the house, so it's time to take our weather station outdoors. You could site the external sensors mast by sticking it firmly into the ground, or securing it to a downpipe with the supplied jubilee clip (gear clamp). Alternatively, as we did, you can use standard zip cable ties to secure the sensor mast to a garden fence post. You could even use a TV aerial mast to mount it on the side of a garage. Ideally, try to avoid siting the weather station too close to trees or other obstructions that may affect the wind and rain readings. Before securing it firmly in place, check Step 8 to calibrate the wind vane.

Since the sensor cables are roughly 3 m in length, you'll need to position Raspberry Pi fairly nearby unless using cable extenders. Naturally, you'll need to protect it from the rain, so you could use a standard weatherproof case or, even better, a Stevenson screen for better ventilation. We simply ran the cables through a window to Raspberry Pi in our garage, although this does give inflated temperature readings on hot days.

06 Test again

With the external sensor assembly complete, and the connectors inserted into the correct ports on the HAT, let's test they're all working correctly.

▲ Choose a block type to add to the dashboard, then select one or more sensor feeds for it

08 Wind vane calibration

To obtain an accurate compass direction reading from the wind vane, you'll need to make sure it's pointing north when its reading is north. To do so, use a standard compass, or an app on your smartphone, to determine in which direction north is.

The wind vane sensor has four barely visible protrusions on its body. When the shorter end of

DOWNLOAD THE FULL CODE:
magpi.cc/weatheraio

the vane is pointing toward the protrusion nearest where the main cable (going to the HAT) comes out, it's pointing north. You can double-check this by running the **weather.py** demo code and rotating the vane so it reads north.

Now rotate the whole mast so that the north end of the wind vane is actually pointing north. Secure it tightly in place so it can't move out of position. You will also want to check that the rain gauge is level, by ensuring that the small spirit level bubble on it is centred.

09 Set up Adafruit IO

While there are numerous options for logging and charting your data locally or online, the easiest way is to use the Adafruit IO code example to set up a web dashboard that you can view from any device.

First, point a web browser to **io.adafruit.com**, click Get Started for Free, and enter your details for a free Adafruit IO account. Click on My Key to see your username and active key. On Raspberry Pi, go to the **examples** folder and edit the code file with:

```
sudo nano adafruit-io.py
```

Now alter the following lines by adding your key and username between the single quote marks:

```
ADAFRUIT_IO_KEY = 'YOUR AIO KEY HERE'
ADAFRUIT_IO_USERNAME = 'YOUR AIO USERNAME
HERE'
```

Press **CTRL+X**, then **Y** to exit and save the file. Then run it with:

```
python adafruit-io.py
```

This will automatically create a Weather dashboard in your Adafruit IO account and populate it with the sensor feeds. Now, you just need to design the web dashboard.

10 Design dashboard

In your Adafruit IO account, open up the weather dashboard. Click the cog icon and Create New Block. Let's start with a temperature graph: select the Line Chart option. In the list of sensor feeds, tick Temperature and then click Next Step. Enter a block title for it, e.g. 'Temperature', leave

the other options unchanged, and click Create Block. It will now appear on the dashboard.

Let's add a real-time temperature gauge. Click the cog and Create New Block, then select the Gauge option. Tick the Temperature feed again and click Next Step, then give it a title, alter the min and max values to your preference (e.g. -20 and 50), and add 'Celsius' as the label. Click Create Block and it'll appear on the dashboard, under the chart.

Using the same process, continue to add dashboard widgets for other sensor feeds, to your preference. We also added a Stream widget to show all the live sensor feeds. You can create a chart for multiple feeds if you like, such as rain and humidity (as we did); you can then select to show/hide each feed by clicking its colour-coded rectangle in the chart key.

To reposition blocks, click the cog icon and Edit Layout, then click and drag the blocks around. You can also click the cog icon on a block to edit it, and alter the history duration for charts. When happy, click Save Layout.

11 Run from boot

With your web dashboard set up and the data uploading correctly, you'll want to get your **adafruit-io.py** script to run automatically on bootup. To do so, enter `crontab -e` and add the following line at the bottom of the file:

```
@reboot python /home/pi/weatherhat-python/
examples/adafruit-io.py &
```

Press **CTRL+X**, then **Y** to save it. Now, even if the power goes off temporarily, when your Raspberry Pi reboots, the script will start running again and continue uploading data. Ⓜ

▲ Use a compass or phone app to ensure the wind vane is aligned correctly to north when reading north

SUPER SIMPLE ROBOTICS

A first foray into the exciting world of robots needn't be scary. There's plenty of help along the way, says **Rosie Hattersley**

I t's nigh on impossible to ignore the rise of the robot; they're everywhere these days: delivering groceries, harvesting our crops, assembling cars, and packaging goods in factories, or entertaining our kids.

Robots are certainly an entertaining form of toy, with plenty of educational value, not least because robots can be taught to learn from us, or the objects we show them, and to react accordingly.

Learning how to code and control a robot is one of the great rewards and challenges of being a Raspberry Pi owner. If you're fairly new to Raspberry Pi, and haven't had much opportunity to experiment with robots or with coding, taking your first step into the world of robots can seem rather daunting. However, there are plenty of kits out there to help you test the waters. Both Raspberry Pi, and the companies that create robot kits, provide step-by-step guidance and support.

" We'd love to hear where your robot adventures take you "

Once you've learned what's involved, and the components you'll need to create a robot of a certain type, you may even feel inspired to design a robot of your own, thus opening the door to a whole new world of creativity and learning! Our tutorial will take you through the excellent CamJam EduKit #3, providing a detailed guide to starting your robot-building journey. We promise that you won't regret giving it a go, and we'd love to hear where your robot adventures take you.

1 WHEELS
Two wheels good, four wheels bad? Or, do you want a super-manoeuvrable castor so your robot can spin and accelerate in any direction? Robots and rovers come in many forms

2 DISTANCE SENSOR
Being speedy is great, but avoiding objects – and other robots – is a must. A distance sensor sends out sonar signals looking for what's in your robot's path and tells it to react or change course accordingly

3 MOTOR
You'll need a motor to drive a rover or robot with wheels. A TT motor like this one works between 3 V and 6 V, making it ideal for Raspberry Pi. Able to spin at different speeds in both directions, two TT motors can drive and turn your robot.

M4GPI 1

THE PARTS THAT MAKE A ROBOT MOVE

Modular parts make for a mighty robot

Designing a robot from scratch involves having a clear vision of what you'd like your creation to be able to do, and breaking down into discrete functions how it's going to achieve each stage. In many cases, you'll find robot kits available that can achieve some or all of what you'd like to accomplish and may be able to adapt to your needs once you've seen how they operate. Working through the assembly process and learning to control a commercially bought kit is useful and rewarding in its own right, and will give you the confidence to go on and design your robot.

Helpfully, many robot kits are Raspberry Pi-based (or have a Raspberry Pi component) and come with, or can be powered by, a Raspberry Pi. There are now Pico microcontroller-based robot parts, such as Kitronik's Motor Driver board for Pico (**magpi.cc/picodriver**).

> ❝ A set of simple TT motors is all you need to get moving ❞

Whether you are taking the plunge and designing a custom robot, or sizing up a suitable kit, these are the essential parts your robot will almost certainly feature.

Raspberry Pi

The sort of robot you want to build will determine the Raspberry Pi version you choose. Key benefits of a Raspberry Pi 4 Model B are its choice of 1GB, 2GB, 4GB, or 8GB configurations and support for AI, a full operating system, plus USB ports. The Pico W board, meanwhile, packs low power into a sleek and stealthy form factor and it powers on, and runs, programs instantly without the dependence of a larger operating system. And Pico W's new-found wireless ability makes it ideal for remote control projects.

magpi.cc/products

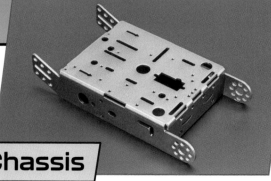

Chassis

Your robot will almost certainly need a body. You can either buy or build (or 3D-print) a chassis. The Pi Hut sells this striking purple chassis, which makes a great body for a two- or four-wheeled robot. It's made from a single piece of 2mm thick aluminium, and has holes and slots for servos, sensors, and mounts. Extra holes can be drilled if required. The precision-engineered robot part is crafted to fit a pair of DC TT motors.

magpi.cc/purplechassis

Motor Controller

Motor controllers offload the speed and direction of your wheels from Raspberry Pi, making it much easier to control movement with code.

For example, this ThunderBorg dual motor controller provides 5 amps of power to two separate motors, making it a great choice for robots that require plenty of oomph! That power is adjustable too, giving you great control over the robot driving or operating experience, including switching into reverse. ThunderBorg is stackable and can be used in multiples for the ultimate in robot power: the MonsterBorg kit gives you an insight into what this controller can do.

`magpi.cc/thunderborg`

Ultrasonic Distance Sensor

How does your robot or rover work out when there's an object it needs to avoid? Using a distance sensor, of course. This one uses twin speaker-like ultrasound transmitters and a receiver to detect obstacles and work out how far away they are. Your code then cleverly instructs the robot to take avoiding action.

`magpi.cc/hrso4`

SG90 Servo

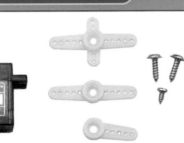

While motors spin, servos rotate back and forth, making them ideal for robotic joints. You'll find them in arms, pinchers, and walking robots. The SG90 is a classic part that you will find in many robot kits. It rotates up to 180º (90º in each direction).

This invaluable part wields plenty of power and connects to many motor controllers via its 3-pin female header.

`magpi.cc/sg90`

TT Motor

This TT motor is an easy-to-use item that you connect to a breadboard via its male jumper leads, or to the aforementioned motor controller to power up and spin around. Dispensing with rotary power dials, this motor offers 200 rpm and between 3 V and 6 V of power (but be aware that it also draws 1.5 V while idling – worth considering if you're making a battery-powered robot, as it doesn't have an 'off' switch). More expensive motors provide higher speed, gear ratios, and encoders that enable precision movement. But a set of simple TT motors is all you need to get moving.

`magpi.cc/ttmotor`

Robot Resources

Make sure that you check out these resources

LEARN ROBOTICS PROGRAMMING 2ND EDITION

Getting to grips with coding in Python will help you get the most from your Raspberry Pi, and will unlock the potential of robotics. This excellent book provides a thorough grounding in how to design and build a robot, then add AI intelligence, lights, and sensors.
▶ magpi.cc/learnrobotics2

ROBOTICS PROJECTS

Robots are one of the coolest ways to use Raspberry Pi – and one of the most popular! Raspberry Pi's walkthroughs show you how to build your own robot buggy and motor, get your robot to follow a line, explore its environment, and control it remotely.
▶ magpi.cc/robotprojects

ROBOTICS WITH RASPBERRY PI

Providing a more pedagogical approach to robotics, Udemy's online course takes you through the essentials of Python coding before using it to control a robot, access on-board cameras, and use robotics in tandem with the Internet of Things and artificial intelligence.
▶ magpi.cc/udemyrobotics

RASPBERRY PI ROBOTICS KITS

Try a self-contained package and build your robot confidence

The CamJam EduKit #3

CamJam's EduKit #3 is an absolute classic kit for anyone wanting to experiment with robots. The kit consists of a motor controller, two DC motors, a distance sensor, plus a ball castor for direction changes, red wheels, and all the jumper leads, connectors, and resistors that you need.

magpi.cc/edukit3

> " CamJam's EduKit #3 is an absolute classic "

You'll love SpiderPi

Kitronik Autonomous Robot

Nifty turns and changes of direction are in the DNA of this Kitronik buggy platform, specially designed for Pico users. With piezo buzzers, lights, and line following capabilities, basing a robot around this buggy will have you absolutely bossing it over the competition.

magpi.cc/kitronikpico

SpiderPi

If your idea of a robot runs around and scares small animals, you'll love SpiderPi, a hulk of flexing metal that responds to visual cues and is user-programmable. It can follow lines on the ground, recognise objects, jump up and down thanks to an inverse kinematic gait, and even pick up and move objects. One of the most detailed robots around.

magpi.cc/spiderpi

Trilobot Base Kit

Sporting some serious bling, including RGB LED under-lighting, this educational but generally awesome robot kit sandwiches most of its electronics between two slices of circuit board that Pimoroni has pimped with an ultrasonic sensor, a built-in camera mount for AI cleverness, and sockets for additions such as STEMMA and Qwiic connectors. Add an SD card, power pack, and Raspberry Pi – or choose the self-contained kit.

magpi.cc/trilobot

Pi-top Robotics

If you're looking for a high-end robot kit for classroom or code club use, this robust setup is a great choice. Featuring more than 50 metal plates, rugged wheels, a camera, and an ultrasonic sensor, plus a range of servo motors, this STEM-focused setup works alongside, or can be bought with, a pi-top 4 Raspberry Pi portable computer.

magpi.cc/pitoprobot

GoPiGo 3

One of the most functional robot kits available for Raspberry Pi. GoPiGo broadcasts its own wireless hotspot, making it easy to connect and the two motors have encoders built-in, making for precision movement. One of the most precise robots you can buy, and it is especially useful for teachers.

magpi.cc/gopigo

BUILD YOUR OWN ROBOT: CAMJAM EDUKIT #3

Robots are a great way to learn physical computing and lots of fun too. Don't know where to start? This is the tutorial for you

MAKER

PJ Evans

PJ is a writer, software engineer, and general tinkerer. He believes no robot is finished until it has googly eyes.

twitter.com/ mrpjevans

Raspberry Pi computers have often been the controller of choice for robot builders. Its small size, combined with the power of Raspberry Pi OS, makes it an ideal choice for simple 'buggy' projects and complex machine-learning autonomous builds alike. By combining Raspberry Pi with a battery, you can make untethered robots that can be controlled by Bluetooth, wireless LAN, or radio. There are literally thousands of options out there, so getting started can be intimidating. We've hand-picked some great beginner kits, and will walk you through them over the next few issues, starting with the CamJam EduKit #3.

01 Get to know your kit

The CamJam EduKit #3 Robotics kit contains everything you need to build your first robot. Inside, you'll find all kinds of bits and bobs. Don't be put off, we'll go through all of

these in time. We're going to start with some basic components and then build on that. Of interest to us are the two motors (which will drive the robot), the big grippy wheels, and the castor ball, which provides the rear 'wheel'. You may be wondering what we're going to mount these on. Well, remove all the components and safely store them, because we're going to use the box itself as the chassis!

02 Prepare the brains!

Our robot, unsurprisingly, is going to be controlled by a Raspberry Pi 4. The kit works with any Raspberry Pi model, so a Zero W would also work well. We recommend using Raspberry Pi OS Lite, but everything will work with the 'regular' OS too. Using Raspberry Pi Imager (**magpi.cc/ imager**), make sure you configure anything you need to access the OS, such as wireless LAN and SSH, using the advanced menu. Now write the image to the SD card and use it to boot your Raspberry Pi. Log in and ensure everything is up-to-date with the following command:

```
sudo apt -y update && sudo apt -y upgrade
```

03 Motor mounting

We're going to begin with the two yellow motors that will drive the wheels. Take the now-empty box, then put aside the blue cover. Flip the larger white portion over so you're looking at the base. At one of the shorter ends (doesn't matter

You'll Need

> CamJam EduKit #3
> **magpi.cc/edukit3**

> 4 × AA batteries

> USB power bank (optional)

▲ You can use the battery pack and Raspberry Pi 4 as 'ballast' at the rear. Add a power bank for true remote computing!

The motors are controlled by this Raspberry Pi 4 using a custom controller HAT

Two motors power the robot forward, directly driving these chunky wheels

which), attach the two motors on either side using the supplied sticky pads. A good tip is to trim one down to just the size of the yellow section, so we have a spare sticky for the next part. The yellow base should fit into the corner on each side, running parallel with the longer box edge, so the black part and the wires are pointing inward, and one white shaft is hanging over the edge of the box.

04 Add some steering

To help guide your little robot on its way, a castor ball has been supplied. This makes handling left and right movements much easier by being able to turn in any direction. You'll need another sticky pad (hopefully you've got a spare from the previous step). Place the sticky pad on the base of the castor ball holder, and mount it on the box base on the other side from the motors, in the middle, a little way in from the edge. Place the ball in it, if not already there. Now, carefully make a small hole next to each motor and feed the wires through. You can now flip the box over.

05 Wiring time

Motors require a lot more current than computers are normally able to safely handle. To solve this problem, CamJam has provided a special

HAT add-on for your Raspberry Pi that allows the motors to be controllable, while powered by a separate power source. Looking at the underside of the board, you'll see three green connectors. You need to connect one motor to the left connector, and the other to the right. Finally, connect the

" The wrong wiring could damage your Raspberry Pi "

battery pack to the centre connector. Take your time with this, as the wrong wiring could damage your Raspberry Pi. Check the diagram for the correct wiring.

06 Connect the controller

If you haven't already, shut down your Raspberry Pi 4 and disconnect it from any power. Place your Raspberry Pi 4 in the box towards the rear. Later on, you can secure it with sticky pads, but it's best to wait until you're happy with the placement and everything has been tested. Connect the controller to the GPIO, being very careful to place it on the correct pins. The controller board should point inward to the Raspberry Pi PCB and, viewing with the USB and Ethernet ports on the right, be placed on the leftmost pins (pin 1

Top Tip

Make a chassis

Why not try and build your own chassis? Maybe you can upscale some junk and make it mobile?

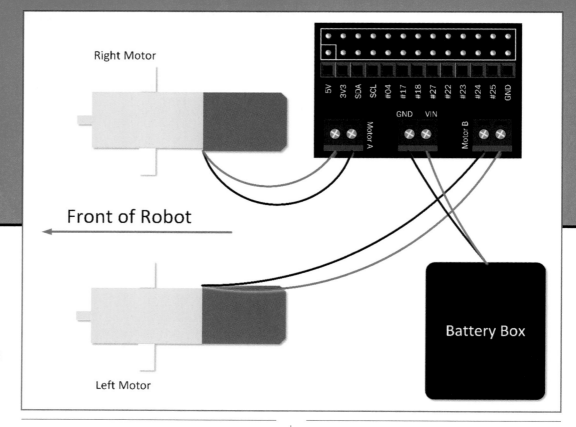

Right Motor

Front of Robot

Battery Box

Left Motor

▶ Wire up the control board as shown here. This is the view looking down

▼ Make sure that the motors are positioned as shown, and that the wheels can attach without scraping the box

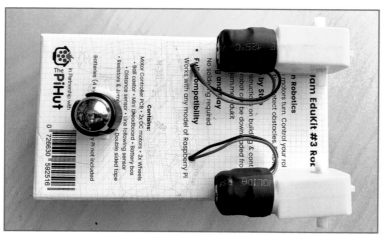

07 Empowering your robot

The supplied battery pack will only serve the two motors, so you need to have a think about how your Raspberry Pi 4 will be powered. If you want to power it from the mains, then the range of your new robotic pal will be limited by the length of cable you can supply. For a true free-range experience, a decent power bank, such as the type that are used to charge phones, is what you need. Make sure it's capable of supplying enough current – a capable output of 3A will be plenty. If you're

08 Power up

If you haven't already, insert four AA batteries into the battery pack and double-check all your wiring from the previous step. Toggle the power switch to 'on'. Absolutely nothing should happen. If anything does, such as a moving motor, switch back to 'off', and check all the wiring. Carefully place the battery box and Raspberry Pi 4 into the chassis box, placing them at the rear, to provide some stability. Now, add your power source for your Raspberry Pi 4 and power up. Don't put the wheels on just yet – it's a lot easier to fix problems if your robot cannot move!

09 Test drive

Let's make sure Raspberry Pi can control the motors. Have a look at the **test_motors.py** listing. These few lines will simply spin up the motors for a second. It uses the GPIO Zero library, which has dedicated commands for controlling this robot, and comes pre-installed with Raspberry Pi OS.

Enter the code using Thonny (or your favourite editor) and save it in your home directory as **test_motors.py**. Now run it:

```
python3 ~/test_motors.py
```

Did both motors spin? If so, you're good to carry on. Otherwise, go back through the steps and check everything again.

10 Back and forth

The next step is to check the motors are correctly wired for backwards and forwards movement. Create a new file in the home directory called **test_direction.py**, and enter the code from the **test_direction.py** listing. This will move the robot forwards, backwards, then left and right. Once you've got the code ready, run it with:

```
python3 ~/test_direction.py
```

If you can't quite see what's going on, try attaching the wheels and see your robot move for the first time! Feel free to play with the code

" You can control the speed "

and try different timings. Just keep an eye on any wires powering Raspberry Pi 4.

11 Make a maze!

Try out your new robot, and give yourself a coding challenge by drawing a maze for the robot to navigate. Create a new Python file, just like before, and use the forward, backward, left, and right commands to create a series of instructions that will drive the robot around the maze. You can control the speed of the robot by adding a value between 0 and 1 after each command. For example, `robot.forward(0.5)` will start the robot moving forward at half-speed. You may find this useful to stop skidding on hard floors.

12 Next month

Congratulations! You now have a small robot pal. If you'd like to find out what else we can do with the CamJam robotics kit, make sure you get next month's issue, where we will have a look at those other components we put aside. Soon, our robot will be able to detect obstacles and follow lines all on its own. Plus, we'll go beyond the kit and look at how you can customise your robot to do even more. Thanks to CamJam for providing assistance with this tutorial! M

test_motors.py

> Language: **Python 3**

magpi.cc/testmotorspy

```
001.  # CamJam EduKit 3 - Robotics
002.  # Motor Test Code
003.
004.  import time  # Import the Time library
005.  from gpiozero import CamJamKitRobot  # Import the CamJam
      GPIO Zero Library
006.
007.  robot = CamJamKitRobot()
008.
009.  # Turn the motors on
010.  robot.forward()
011.
012.  # Wait for 1 seconds
013.  time.sleep(1)
014.
015.  # Turn the motors off
016.  robot.stop()
```

test_direction.py

DOWNLOAD THE FULL CODE:

> Language: **Python 3**

magpi.cc/testdirectionpy

```
001.  # CamJam EduKit 3 - Robotics
002.  # Driving and Turning
003.
004.  import time  # Import the Time library
005.  from gpiozero import CamJamKitRobot  # Import the CamJam
      GPIO Zero Library
006.
007.  robot = CamJamKitRobot()
008.
009.  robot.forward()
010.  time.sleep(1)  # Pause for 1 second
011.
012.  robot.backward()
013.  time.sleep(1)
014.
015.  robot.left()
016.  time.sleep(0.5)  # Pause for half a second
017.
018.  robot.right()
019.  time.sleep(0.5)
020.
021.  robot.stop()
```

Part 02

Build a robot:
add sensors to the chassis

Last month, we started our build of the CamJam Robotics EduKit.
Now we have a roving robot, it's time to add some smarts!

MAKER

PJ Evans

PJ is a writer, software engineer and tinkerer. His robots bring all the geeks to the yard.

twitter.com/ mrpjevans

I f you followed last month's tutorial, you should now have a working robot that you can control with Python. Hopefully, you've played with the code and had the little 'bot' zooming around the place. Now it's time to add some sensors, so our new pal can start to sense the world around it. With the ultrasonic and light sensors included with the CamJam EduKit #3, we can add some autonomous capabilities and make our robot a little smarter. Finally, we can look at what you can do to improve the robot even more with custom chassis and additional sensors.

01 Get sensitive

Included with your CamJam kit is a light sensor. It works by sending out infrared light (that we can't see) and detecting how much of it bounces back to a sensor. If we point it at the ground and measure the sensor's output, we can easily tell when the robot passes over a line. The key to success is high contrast, so a jet black line on a white surface is perfect. We're going to mount the sensor on the front of the robot, point down, so we can detect a line.

You'll Need

> CamJam Edukit #3 - Robotics **magpi.cc/edukit3**

> Printer

> Roll of paper (optional)

02 A little light wiring

To wire up the light sensor, we're going to use the breadboard (the small block with lots of holes). Holding it with the longer edge horizontal, each column of holes are connected together, with a gap in the centre. Breadboards allow us

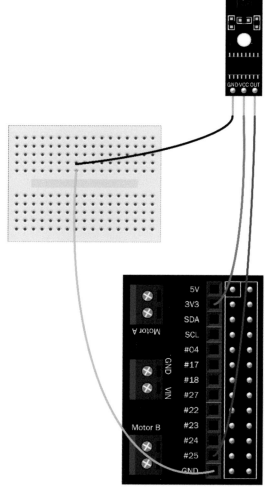

▲ **Figure 1** Wire up the line sensor to the HAT. Use the breadboard to create a 'ground rail'

By measuring how long it takes to receive an ultrasound echo from the transmitter, we can calculate how far the robot is from an obstacle

This tiny sensor bounces light from a transmitter to a receiver. It can detect lines through a sudden drop in reflected light

to connect circuits together without soldering, so we can quickly prototype circuits and correct mistakes. You'll need three wires: two plug-to-socket, and one plug-to-plug. Wire everything together as shown in the diagram, checking and double-checking. There are three connectors on the light sensor for power, grounding, and data. These all need to match up with the connector on the robot HAT connected to your Raspberry Pi.

03 Mount the sensor

The sensor needs to be in a sensible place on the box chassis, and that would be in the centre at the front on the base. However, this is also the highest point of the body, and the further away the sensor is from the ground, the less accurate it will become as ambient light leaks into the sensor. Start by making a hole off-centre towards the front of the chassis and feeding the three wires through it. Connect the wires to the sensor as shown in **Figure 1**, then mount the sensor to the body with sticky pads. You may find a couple of LEGO bricks will sufficiently lower the sensor to the ground.

04 Testing time

The sensor will not work without a little code to help it on its way. Enter the code from the **test_line.py** listing, overleaf (or download it from the GitHub repo: **magpi.cc/testlinepy**), and save it in your home directory as **test_line.py**. Now run the code:

```
python3 test_line.py
```

Don't worry, your robot won't move at this point. What we want to do is check the sensor is working correctly. Using a sheet of paper with a thick black line through it (**magpi.cc/testlinepdf**), hold your robot carefully and pass the paper under the sensor. If all is working well, you'll see messages on-screen that the line has been detected.

05 Follow that line

Now our little pal can detect a line, we can make them follow it too! By combining code to drive the robot forward and steer it left and right, we can make corrections as we go. This listing is a

Top Tip

Safety First

Whenever connecting wires to your Raspberry Pi computer, ensure it is completely powered-down. A mistake when the computer is on can cause permanent damage.

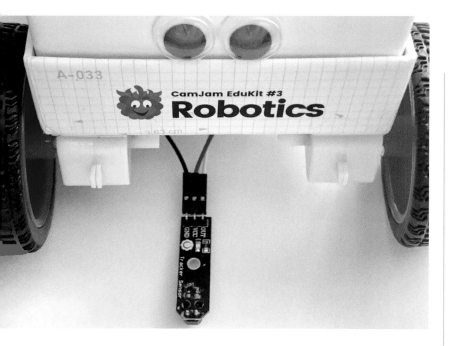

Here is the wiring on the line sensor. The lower the sensor can be mounted to the ground, the more accurate the result will be

little longer, so download it to your home directory from **magpi.cc/linefollowerpy** and try it out. Place your robot at the start of the line and then run:

```
python3 line_follower.py
```

Hopefully, your roving friend will scoot along the length of the page. Do you notice how it's slower and more controlled? We're using pulse-width modulation (PWM) to slow the motors down. You can play with the setting by changing the `leftmotorspeed` and `rightmotorspeed` variables.

06 Make a course, of course

OK, so our new friend can follow a line. How about an entire course? If you've got access to a big roll of white paper, try mapping out a course for the robot to follow. If not, you could stick pieces of A4 paper together. Make it as big as you can, without any tight corners to which the robot may not be able to respond. Use a pen such as a Sharpie to create the line to follow, and make it nice and thick, like the one on the printout. Now watch as your newly smarter robot follows the line in circles.

07 Looking into the distance

If we want our robot to be able to move around a room on its own, there's a significant problem: walls. Our final modification is to add a distance sensor to the robot, so it can take avoiding action when it gets close to an obstacle.

Top Tip

Colourful resistors

Resistors are identified by coloured stripes on their body and can be used either way around. The 470 Ω resistor is Yellow, Violet, Black, Black, Brown; the 330 Ω is Orange, Orange, Black, Black, Brown.

The sensor works by transmitting an ultrasonic pulse and detecting when it is returned. With a bit of maths, we can use the time taken ('time of flight') to calculate how far away the obstacle is from the robot. The wiring is a little more complicated for this as the sensor needs 5V to work, but must only return a 3.3V signal to avoid damaging your Raspberry Pi 4.

08 Wiring up for safe voltage

Study the **Figure 2** wiring diagram carefully. Mount the sensor to the breadboard along the long edge, so each connector has its own column (or 'rail'). Move the two existing ground connectors for the line sensor so they are on the same rail as the GND pin for the distance sensor. Connect TRIG to #17 on the CamJam HAT. Finally create a 'voltage divider' to reduce the return voltage to 3.3V. Do this by adding the supplied 470 Ω resistor to bridge the GND rail to any spare rail. Now add the 330 Ω resistor to bridge that spare rail to ECHO. Finally, link the the spare rail to #18 on the HAT.

test_line.py

DOWNLOAD THE FULL CODE: magpi.cc/testlinepy

> Language: **Python 3**

```python
001. import time
002. from gpiozero import LineSensor
003.
004. # Set variables for the GPIO pins
005. pinLineFollower = 25
006. sensor = LineSensor(pinLineFollower)
007.
008. def lineseen():
009.     print("Line seen")
010. def linenotseen():
011.     print("No line seen")
012.
013. # Tell the program what to do with a
     line is (un)seen
014. sensor.when_line = lineseen
015. sensor.when_no_line = linenotseen
016.
017. # Repeat the next indented block
     forever
018. while True:
019.     time.sleep(10)
```

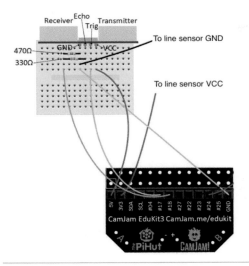

09 Facing forward

To get an accurate reading, the ultrasonic sensor needs to be mounted facing forward in the centre. You may have to get a little creative to find the best way to attach the breadboard so it fits. We used a bit of double-sided sticky tape on the breadboard to hold in place, so the sensor sat over the edge of the box. A small cardboard or plastic box for the board to sit on would also work well. Another option is to remove the sensor from the breadboard altogether and use four jumper wires to reconnect it, allowing the breadboard to sit on the base.

10 Testing from a distance

Let's create another test file. In your home directory, create a new file called **test_distance.py** and add the code from the listing here (or download it from **magpi.cc/testdistancepy**). As before, run this code from the command line:

```
python3 test_distance.py
```

Watch the output from the screen and move your hand towards the sensor. If everything is working, you'll see measurements of the distance from your hand to the robot. This is calculated by taking the output from the sensor (the elapsed echo time in seconds), multiplying it by the speed of sound (34,326 cm per second) and then halving, as it has made an outward and return journey.

11 You own autonomous robot

Congratulations! Your robot build is now complete. Let's combine all the parts of the robot in one last Python program. It's a bit long, so you can download it here: **magpi.cc/avoidancepy**. The code will move the robot forward until it detects an upcoming obstacle. It will then back off and turn right. It will then advance forward. If the obstacle is still in place (or a new one is found), it will repeat the process until the obstacle is cleared. These simple instructions will result in a robot that will happily trundle around the room until you stop it or the batteries wear out!

12 Make it your own

You are the proud custodian of a line-following, obstacle-detecting robot. The good news is, that's not the end of your, or your robot's, journey. Now you have the basic building blocks of code, try and make your robot do more. Could you add dynamic speed control based on distance from an obstacle? Are there other sensors you could add? How about 3D-printing Daniel Bull's custom-designed chassis (**magpi.cc/robotchassis**)? Or, use this project as a starting point and design and build your own robot. Try replacing the CamJam HAT with a motor controller board and upgrading to four-wheel drive. Whatever you decide, have fun.

Many thanks to Mike Horne and Tim Richardson of CamJam for their help sheets and code that informed this tutorial. 🅜

◄ **Figure 2** Build this circuit to get readings from the distance sensor. Always triple-check everything before powering-up!

---

test_distance.py

DOWNLOAD THE FULL CODE:

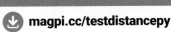
magpi.cc/testdistancepy

> Language: **Python 3**

```
001.  import time
002.  from gpiozero import DistanceSensor
003.
004.  # Define GPIO pins to use on the Pi
005.  pintrigger = 17
006.  pinecho = 18
007.  sensor = DistanceSensor(echo=pinecho,
          trigger=pintrigger)
008.
009.  # Check the distance every half a second
010.  while True:
011.      print("Distance: %.1f cm" % (
          sensor.distance * 100))
012.      time.sleep(0.5)
```

Build a 64-bit
Minecraft server

The latest Minecraft update brings great new features, but also new system requirements. Use 64-bit Raspberry Pi OS to set up a multiplayer server

MAKER

PJ Evans

PJ is a writer, tinkerer, and software engineer. His son made him write this tutorial and has been online ever since.

twitter.com/ mrpjevans

Minecraft has been a feature of the Raspberry Pi ecosystem since the early days. Although the game is demanding on resources, a cut-down educational version became available and was very popular. As Minecraft has developed with new features, challenges, and mods, the 32-bit world was left behind – especially as the most recent release was 64-bit only, owing to the version of Java required. With the release of Raspberry Pi OS 64-bit, the little computer that can is back in the game with all the power of the latest Raspberry Pi 4 too. We're going to show you how to build a Minecraft multiplayer server for you and your friends to enjoy.

01 Set up your hardware

Setting up a Minecraft server may not seem like a hardware project, but the performance of the server is critical to fun gameplay. Firstly, we need the best Raspberry Pi can offer, and that's a Raspberry Pi 4. You can use an earlier model if you wish, but you won't get the smooth gameplay that the latest and greatest can offer you. A big factor is memory. Java uses a lot of memory and Minecraft needs still more. If you can, get an 8GB model, 4GB minimum. Finally, avoid wireless LAN if at all possible. Use a hardwired connection to your router if you can.

You'll Need

▸ Raspberry Pi 4 case with cooling
magpi.cc/casefan

▸ Minecraft Java Edition
minecraft.net

▸ Spare router port (optional)

▸ Ethernet cable (optional)

▸ M.2 SSD drive (optional)

02 Prepare your SD card

We've suggested using a USB-boot M.2 SSD drive for best performance, but you can use a regular SD card too. Just make sure it's high quality as it will be heavily used by Minecraft. Either way, the process is the same: using Raspberry Pi Imager

Network lag is the enemy of good gameplay, so avoid wireless LAN and use a wired connection to your router

The Minecraft server puts a lot of strain on a Raspberry Pi 4B, so a good case with built-in cooling is essential

(**magpi.cc/imager**) select Choose OS > Raspberry Pi OS (other) > Raspberry Pi OS Lite (64-bit). Now select your card under 'Storage', then click the cog icon. Set the hostname (we used 'steve'), enable SSH, set a password for the user 'pi', and configure your wireless LAN (if using). Now write your image.

03 Install dependencies

A dependency is a piece of software that is needed by the software we want to run. In our case, this is Java, a popular programming language and platform that was used to write Minecraft. The latest version of the Minecraft server requires version 17 or above, which is only available as 64-bit. Before installing, log into the OS and make sure everything is up-to-date with:

```
sudo apt -y update && sudo apt -y upgrade
```

Once complete, install the Java runtime:

```
sudo apt install openjdk-17-jdk git
```

(We're also installing Git, as we'll need it later). Test everything has worked:

```
java --version
```

If you get the version number back, you're good to go.

> ## ❝ The performance of the server is critical to fun gameplay ❞

04 Create a dedicated user

As we may want others to join our Minecraft games, it makes sense, and good practice, to run Minecraft under a dedicated user account with restricted permissions. That way, if someone gains access to the server via Minecraft, they will be heavily restricted on what they can do. To create the user:

```
sudo useradd -r -m -U -d /usr/local/
minecraft -s /bin/bash minecraft
```

We now need to assume that user's identity to continue. To do this we can 'swap user' or 'su':

```
sudo su - minecraft
```

Now, we'll create some local directories to store the things we need:

```
mkdir -p ~/{backups,tools,server}
```

This neat trick allows us to create several directories with one command.

▲ Start Minecraft Java Edition and then select Multiplayer from the main menu

05 Build the console tool

Minecraft comes with a control interface called rcon. So that Minecraft can be shut down tidily when we shut down the computer, we need to ask Minecraft to save and stop, using rcon. A neat utility called mcrcon helps us do this. We need to download the source code and build it, so let's start with downloading the code from GitHub. As the 'minecraft' user:

```
cd ~/tools
git clone https://github.com/Tiiffi/mcrcon.
git
```

Now we can build the utility:

```
cd ~/tools/mcrcon
gcc -std=gnu11 -pedantic -Wall -Wextra -O2
-s -o mcrcon mcrcon.c
```

To check it works, run the help screen:

```
./mcrcon -h
```

06 Download the Minecraft server

We now have everything we need to install and run Minecraft. It's time to download the server and set it up. To find the latest version, visit **magpi.cc/mcserver**, right-click on the download link, and 'Copy link'. Using that link, and as the 'minecraft' user, download the server directly into the **~/server** directory on Raspberry Pi 4:

Top Tip

Get an M.2 boost

For a speed boost, replace the SD card with an M.2 SSD. Cases such as the Argon M.2 support them as a USB boot device.

▲ You're online in a virtual world with your friends

```
cd ~/server
wget https://launcher.mojang.com/v1/objects/
e00c4052dac1d59a1188b2aa9d5a87113aaf1122/
server.jar
```

Note: this link is the latest – v1.19 – at the time of writing.

Now, run the server for the first time. It will configure a lot of things and then exit:

```
java -Xmx1024M -Xms512M -jar server.jar
nogui
```

Top Tip 👍

VPN

An alternative to having an open Minecraft server port is to install a VPN server such as WireGuard (piVPN) and allow friends to connect to it.

07 Configure the server

Before you can go around chasing creepers, we need to set a few things up on the server. You may have spotted a message about the end–user licence agreement (EULA) on the previous run. We need to set that (to show we agree) and change a couple of things to enable the rcon protocol.

▶ If your client can't find the server, don't worry, just select 'Direct Connection'

```
nano ~/server/eula.txt
```

Change false to true, then save and exit (**CTRL+X**).

```
nano ~/server/server.properties
```

Find `rcon.password=` and add a strong password of your choosing (remember it!), then find `enable-rcon=` and change `false` to `true`. Now save and exit.

08 Test it!

Let's try it out. The command to start Minecraft manually is:

```
java -Xmx4096M -Xms512M -jar server.jar
nogui
```

The number after `-Xmx` is the maximum amount of memory Minecraft can use. Here, it is 4GB (4096MB). Start with this and then try increasing it by 1GB at a time if you run into trouble.

Run up the server using the command above and, once running, see if you can connect to it using Minecraft Java Edition on your device of choice. Click on 'Multiplayer' then 'Direct Connection' then enter the server name, in our case 'steve.local'. After a few seconds, you should be dropped into the world and you're online! **CTRL+C** stops the server.

09 Run as a service

Ideally, we'd like the Minecraft server to run on boot, so we don't have to log in and start it manually. We can create a system service to do this. Start by creating the configuration file:

```
exit # If you're still the minecraft user
sudo nano /etc/systemd/system/minecraft.
service
```

Add the contents of the **minecraft.service** listing on this page (or copy from **magpi.cc/minecraftservice**). Remember to change `password` to the actual password you thought of in the previous step. Now enable the service:

```
sudo systemctl enable minecraft
```

▲ Enter the hostname of your server followed by '.local'. If joining from outside the network, this needs to be your router IP address

Finally, `sudo reboot` to restart the OS and you can check that Minecraft is running by checking its status:

```
sudo systemctl status minecraft
```

❝ Investigate services such as DynDNS ❞

10 Configure your router

Your Minecraft server is now up and running, but no one outside of your home can access it! To allow friends and family to play along, we need to ask your home router to permit access. This is something to be done cautiously, as you do not want to accidentally allow people access to parts of your network. Consult the instructions for your router, and follow these steps: first, set up a DHCP reservation for your Minecraft server; this means it will never change IP address. Secondly, set up port forwarding from port 25565 on the router to the same port on your server. You may need to restart your router for the changes to take effect.

11 Testing access

For others outside your network to play, you'll need to give them your public IP address (your router will tell you this). It may take a couple of goes to get access sorted out. One issue you may come across is your router having a dynamic IP address, so occasionally the IP address

it presents to the world changes to something new. If this becomes a pain, investigate services such as DynDNS that assign a domain to your IP which updates automatically if your IP changes. Now, an external user should be able to join your server using your public IP address.

12 Securing access

You now have a publicly available Minecraft server. That might not be a good idea! To stop anyone, including strangers, joining, we recommend implementing the whitelist feature. This will allow only named players (i.e. people you know) from joining. In Minecraft itself, you can enable the whitelist with `/whitelist on` and then control access using `/whitelist add <user>` and `/whitelist remove <user>`. This is an essential step to keep your server safe. Also, consider changing the number of your external port to something else, as this can help dodge bots scanning for Minecraft servers. Have fun!

Thanks to Heidy Ramirez of **shells.com** for their excellent blog post on this subject.

---

minecraft.service

DOWNLOAD THE FULL CODE:
⬇ magpi.cc/minecraftservice

> Language: **Shell**

```
001. [Unit]
002. Description=Minecraft Server
003. After=network.target
004.
005. [Service]
006. User=minecraft
007. Nice=1
008. KillMode=none
009. SuccessExitStatus=0 1
010. ProtectHome=true
011. ProtectSystem=full
012. PrivateDevices=true
013. NoNewPrivileges=true
014. WorkingDirectory=/usr/local/minecraft/server
015. ExecStart=/usr/bin/java -Xmx4096M -Xms512M -jar
     server.jar nogui
016. ExecStop=/usr/local/minecraft/tools/mcrcon/mcrcon -H
     127.0.0.1 -P 25575 -p password stop
017.
018. [Install]
019. WantedBy=multi-user.target
```

This strip of LED lights and a mirror tile are used to create the visual effect

Raspberry Pi Pico controls the LED strip and takes centre stage in the project

Raspberry Pi Pico
Iron Man Arc Reactor

Build your very own Arc Reactor prop from Iron Man using LED lights

MAKER

Toby Roberts

Raspberry Pi Maker in Residence.

raspberrypi.com

If you enjoy Raspberry Pi-based DIY build projects, there's a good chance that Iron Man might just be one of your favourite superhero characters. A billionaire inventor who created a suit of armour powered by a small, powerful electric generator known as an Arc Reactor — what's not to like? We're going to build our own Arc Reactor using a strip of LEDs and some wizardry to produce a 3D infinity mirror effect.

01 How does it work?

In this tutorial, we'll use a Raspberry Pi Pico to control 31 individually addressable LED lights mounted between two discs of acrylic plastic. One of those discs will have a layer of

adhesive mirror sheet and the other will have a one-way mirror film; this will give the LEDs a 3D infinity effect. Unfortunately, we haven't yet perfected our own plasma fusion power source for Raspberry Pi, so instead, we will use a rechargeable battery and enclose everything in a 3D-printed case.

02 MicroPython

Raspberry Pi has extensive documentation for Raspberry Pi Pico (**magpi.cc/picodocs**), but in this first step we will flash the Pico with firmware using the incredibly user-friendly, drag-and-drop method to transfer files to it (just as you would transfer files to a USB memory stick).

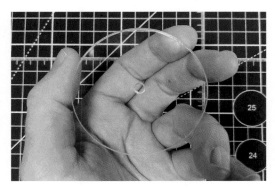

▲ **Figure 1** These acrylic discs can be created using a laser cutter, or cut manually using a jigsaw and sanding paper

On your computer, download the UF2 file for the latest release of the Pico MicroPython firmware from here. The MicroPython programming language is an implementation of Python that's optimised for use with microcontrollers, and it's considered to be one of the best languages for programmers of all levels of experience.

> # " Using a strip of LEDs and some wizardry to produce a 3D infinity mirror effect "

03 Installing the firmware

To copy the UF2 file you've just downloaded over to your Pico, you'll need first to put it into bootloader mode. To do this, hold down the BOOTSEL button (the small button next to the USB port) while simultaneously plugging a micro USB cable connected to your Pico into your computer. Your Pico should now show up as a drive called 'RPI-RP2:'.

Locate the .uf2 firmware file that you just downloaded, and drag and drop it into the RPI-RP2 drive, or simply copy and paste it over. Your Pico will now reboot automatically. Once you've done this, your Pico won't show up as a drive again when it's plugged in, but keep it connected ready for the next step. That's all there is to it – you've flashed the firmware.

04 Programming your Pico

Download, install, and open Thonny (**magpi.cc/thonny**), a Python IDE (integrated development environment). It's the software you'll use on your computer to program your Pico when it is connected via USB.

If you see >>> in the Shell window, then you're already connected to your Pico and have an interactive session enabled, and you're ready to move on to programming your Pico. If you don't see this, you need to check that Thonny is set up correctly. Click on the very bottom-right corner of the Thonny window to make sure that the MicroPython (Raspberry Pi Pico) interpreter is selected; if it isn't, select it. If for some reason you weren't successful in flashing the firmware, Thonny may prompt you to install it at this stage; in this case, try flashing it again. If your Pico is still not showing up as connected, disconnect and reconnect it, then press the red STOP sign in Thonny's top menu bar to reset everything. You should now see the >>> prompt in the Shell window, meaning you're connected and ready to continue.

Now you're ready to program your Pico. Enter the code from **main.py** (overleaf). Or copy and paste the code from **magpi.cc/arcreactor** into the empty, and currently untitled, Thonny program window.

05 Read the code

One of the advantages of using MicroPython is that much of it is written in readable English. For example, at the beginning of this program, we can see that Pico will control 31 LEDs via pin 28, and that they are to be set at maximum brightness (1 on a 0–1 scale, 0.5 being 50% brightness). The rest of the program instructs your Pico to display patterns and colours on the LEDs repeatedly.

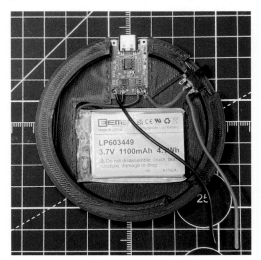

◄ **Figure 2** Soldering the electronics into the USB charging board

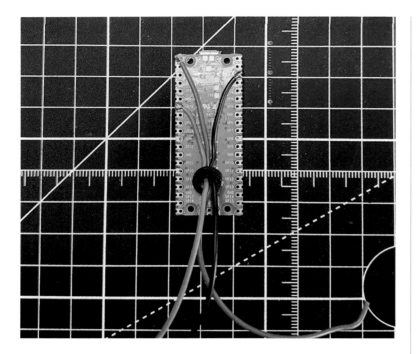

▲ **Figure 3** Soldering the wires to Raspberry Pi Pico. Pico is shown here with the wires threaded through the 3D-printed stand

Click on File and Save as... and a pop-up will appear prompting you to specify where you want to save the file. Click on Raspberry Pi Pico and name the file 'main.py'.

It's important to name the file **main.py** because any file with this name will be started automatically every time Pico is powered on. That's all there is to it: you've programmed your Pico.

Warning! CNC Cutter

Be careful when using CNC cutters in your projects. Wear ear protection and safety glasses and stand clear of the machinery as it works. Understand the basic safety functions of your machine.

magpi.cc/cncsafety

06 The 3D-printed parts

Four 3D-printed pieces are required to house all the component parts: the back, main body, Pico stand, and front. We will simply glue them together to form the complete assembly. You can download the 3D print files free from Printables (**magpi.cc/arcreactorstl**). Print them using any FDM (fused deposition modelling) 3D printer with a build area of 70 mm🅇 or larger; we recommend you use a material that is easy to print with, such as PLA or PETG filament.

07 Cutting and preparing the acrylic discs

For this project, we will need two acrylic discs 3 mm thick and 70 mm in diameter, one of which will require a hole approximately 5 mm in diameter

for wiring. If you have access to a laser cutter, making these is straightforward. Otherwise, it might be time to dig out your old pencil case and find a pair of compasses. Using a jigsaw, cut out a 70 mm circle, and neaten up the edges with sandpaper or a file. The final discs don't need to be absolutely flawless, since the subsequent steps will cover any minor imperfections. One disc should be drilled with a 5 mm hole in the centre to allow wiring to pass through later (see **Figure 1**).

08 Prepare the mirror

Mark a 70 mm circle on your flexible adhesive mirror tile, and another on your one-way mirror self-adhesive film. It should be fairly easy to get them perfectly round by cutting them out using scissors. Ensure that you remove all the protective layers from your acrylic discs, and then peel the adhesive backing from your mirror sheets in turn. Attach the circle of the mirror tile sheet to the disc with the hole, which will be used to mount your Pico, and the circle of one-way mirror film to the other disc.

> ❝ Attach the circle of the mirror tile sheet to the disc with the hole ❞

09 Wiring and soldering the electronics

The Arc Reactor base will contain the rechargeable battery, on/off switch, and USB-C charging board, all of which you need to glue into place inside the 3D-printed cutouts. At this stage, you need to do some cutting of wires, stripping, and soldering. Use **Figure 2** as a reference to ensure that the positive and negative wires from the battery are soldered to the correct USB-C charging board positive and negative inputs and that the positive output from the board is soldered to the centre pin

Top Tip

YouTube Video

Take a look at the Arc Reactor video on YouTube to see how it all works: **magpi.cc/arcreactoryt**.

main.py

> Language: **Python**

```python
001.    import array, time
002.    from machine import Pin
003.    import rp2
004.
005.    # Configure the number of WS2812 LEDs.
006.    NUM_LEDS = 31
007.    PIN_NUM = 28
008.    brightness = 1
009.
010.    @rp2.asm_pio(sideset_init=rp2.PIO.OUT_LOW,
            out_shiftdir=rp2.PIO.SHIFT_LEFT, autopull=True,
            pull_thresh=24)
011.    def ws2812():
012.        T1 = 2
013.        T2 = 5
014.        T3 = 3
015.        wrap_target()
016.        label("bitloop")
017.        out(x, 1)              .side(0)    [T3 - 1]
018.        jmp(not_x, "do_zero")  .side(1)    [T1 - 1]
019.        jmp("bitloop")         .side(1)    [T2 - 1]
020.        label("do_zero")
021.        nop()                  .side(0)    [T2 - 1]
022.        wrap()
023.
024.    # Create the StateMachine with the ws2812 program,
            outputting on pin
025.    sm = rp2.StateMachine(0, ws2812, freq=8_000_000,
            sideset_base=Pin(PIN_NUM))
026.
027.    # Start the StateMachine, it will wait for data on its
            FIFO.
028.    sm.active(1)
029.
030.    # Display a pattern on the LEDs via an array of LED RGB
            values.
031.    ar = array.array("I", [0 for _ in range(NUM_LEDS)])
032.
033.    ########################################################
            ###################
034.    def pixels_show():
035.        dimmer_ar = array.array("I", [0 for _ in range(
            NUM_LEDS)])
036.        for i,c in enumerate(ar):
037.            r = int(((c >> 8) & 0xFF) * brightness)
038.            g = int(((c >> 16) & 0xFF) * brightness)
039.            b = int((c & 0xFF) * brightness)
040.            dimmer_ar[i] = (g<<16) + (r<<8) + b
041.        sm.put(dimmer_ar, 8)
042.        time.sleep_ms(10)
043.
044.    def pixels_set(i, color):
045.        ar[i] = (color[1]<<16) + (color[0]<<8) + color[2]
046.
047.    def pixels_fill(color):
048.        for i in range(len(ar)):
049.            pixels_set(i, color)
050.
051.    def color_chase(color, wait):
052.        for i in range(NUM_LEDS):
053.            pixels_set(i, color)
054.            time.sleep(wait)
055.            pixels_show()
056.        time.sleep(0.2)
057.
058.    def wheel(pos):
059.        # Input a value 0 to 255 to get a color value.
060.        # The colours are a transition r - g - b - back to r.
061.        if pos < 0 or pos > 255:
062.            return (0, 0, 0)
063.        if pos < 85:
064.            return (255 - pos * 3, pos * 3, 0)
065.        if pos < 170:
066.            pos -= 85
067.            return (0, 255 - pos * 3, pos * 3)
068.        pos -= 170
069.        return (pos * 3, 0, 255 - pos * 3)
070.
071.    def rainbow_cycle(wait):
072.        for j in range(255):
073.            for i in range(NUM_LEDS):
074.                rc_index = (i * 256 // NUM_LEDS) + j
075.                pixels_set(i, wheel(rc_index & 255))
076.            pixels_show()
077.            time.sleep(wait)
078.
079.    BLACK = (0, 0, 0)
080.    RED = (255, 0, 0)
081.    YELLOW = (255, 150, 0)
082.    GREEN = (0, 255, 0)
083.    CYAN = (0, 255, 255)
084.    BLUE = (0, 0, 255)
085.    PURPLE = (180, 0, 255)
086.    WHITE = (255, 255, 255)
087.    COLORS = (
            BLACK, RED, YELLOW, GREEN, CYAN, BLUE, PURPLE, WHITE)
088.
089.    while True:
090.
091.        print("fills")
092.        for color in COLORS:
093.            pixels_fill(color)
094.            pixels_show()
095.            time.sleep(0.2)
096.
097.        print("chases")
098.        for color in COLORS:
099.            color_chase(color, 0.01)
100.
101.        print("rainbow")
102.        rainbow_cycle(0)
```

of the sliding switch. The positive wire from the switch can be soldered to either of the two outer switch terminals.

10 Solder to Pico

The next step is to solder three wires directly to the back of your Pico. The wires should be long enough to complete the wiring circuit later in the assembly process: approximately 20 cm should be sufficient. In order to provide power to Pico, you need to connect red and black wires to the pins marked VBUS and GND respectively. We need a third wire, shown in blue in **Figure 3**, that you should solder to the pin marked GP28. Our MicroPython script specifies this as the pin used by Pico to communicate with the LEDs.

11 Wiring loom

LED strips are usually pre-wired, but their joints are often bulky, so we will make our own wiring loom. Using a pair of scissors, remove any pre-existing wiring and cut a strip of 31 LEDs, making sure to cut along the line between each LED (see **Figure 4**).

The strip is also marked with arrows to show the correct current direction, + symbols for the positive wire, 0 for the data wire, and G for the

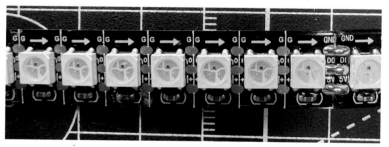

▲ **Figure 4** Cutting the LED strip down to the right size

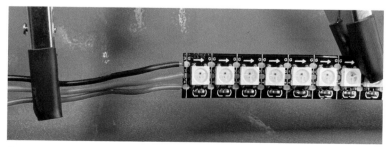

▲ **Figure 5** Soldering the wires onto the LED strip

Top Tip

Test, then glue

Don't forget to test everything works before gluing it all together.

negative or ground wire. It's important, when cutting, to make sure that you snip down the middle of each solder pad; if you're not careful, it makes it tricky to solder the wires to the pads.

12 Solder more wires

Solder three more wires, also approximately 20 cm in length, as shown in **Figure 5**: red for positive, blue for data, and black for ground. When soldering to the pads, you may find it more convenient to do so from the rear of the strip.

> ❝ Ensure that Pico sits above the surface of the mirror on the stand ❞

13 Assembly

Feed the three wires attached to your Raspberry Pi Pico through the small 3D-printed Pico stand. Then thread the wires through the hole in the mirrored disc, and glue the stand to the underside of your Pico and to the reflective side of the disc. Ensure that Pico sits above the surface of the mirror on the stand. In doing so, we will achieve our goal of a 3D infinity effect (see **Figure 6**).

14 Glue and stick

Now, place the disc with the one-way film in the main 3D-printed body and glue the front ring to the body. Because the disc is held in place by the front ring, any minor imperfections in its shape will be concealed.

Stick the strip of 31 LEDs around the inside of the 3D-printed body, ensuring that the wiring and connections are aligned with the gap in the body, so that you can easily pass the wires around the side of the disc on which your Pico is mounted.

◀ **Figure 6** Attaching the Raspberry Pi to the mirror disc with glue

▼ **Figure 7** How everything fits together

Most LED strips have a self-adhesive backing, which helps make this straightforward. Refer to the diagram in **Figure 7** to see how everything will fit together.

15 Solder it all together

With Pico already glued to its mirrored disc, you can now pair it with the main body housing the LEDs and the one-way mirror, and with the base containing the battery, charging board, and switch. Make sure that you have the ends of all your wires threaded through to the Arc Reactor base. Solder the two blue data wires together, all three red positive wires together, and all three negative ground wires together, trimming any excess wire length as necessary. Don't forget to insulate your joins with heat-shrink tubing or tape.

16 Final checks

Before you glue the pieces together and make it final, check that everything is working as expected and that your LEDs are lighting up by sliding the switch. Make sure the charging board functions by attaching a USB-C phone charger or a USB battery pack; a small LED will illuminate when it's charging.

Now all that's left to do is to glue all the parts together. ▨

RETRO GAMING
with Raspberry Pi Pico and Pico W

What kind of video games can you run on a $6 microcontroller?

WRITER

K.G. Orphanides

K.G. is a software preservationist and developer with an abiding love of vintage computers.

magpi.cc/MightyOwlbear

From Dizzy to Doom, Raspberry Pi Pico and Pico W are holding their own against far more expensive emulation systems, and they're ready to play almost as soon as you power on. With the July 2022 launch of the Raspberry Pi Pico W, equipped with wireless networking capabilities, there are a few more features to be tapped.

Numerous mature and highly capable emulators have been ported to Pico, allowing you to run huge chunks of the 8-bit computer and console eras, with hardware kits and expansion boards that make it easy to connect monitors, controllers, SD card storage, and high-quality audio.

Unlike top-heavy computer emulators designed for computers and capable of running almost anything, if you spend enough time poking at the menus, RP2040 emulators become embedded systems, single-purpose, responsive, and fast-booting, just like the original hardware they're emulating.

Emulation is all well and good, but you can also use your Pico to bring brand new homebrew games to real retro hardware platforms, giving new life to beloved classic consoles. We've showcased two different flash carts, for the Nintendo Game Boy and N64 respectively, that use Pico with custom PCBs, that allow you to write your own retro console cartridges at a far lower cost than dedicated commercial flash carts.

▲ Play indie gems from modern C64 developers and publishers, like Psytronik's Honey Bee

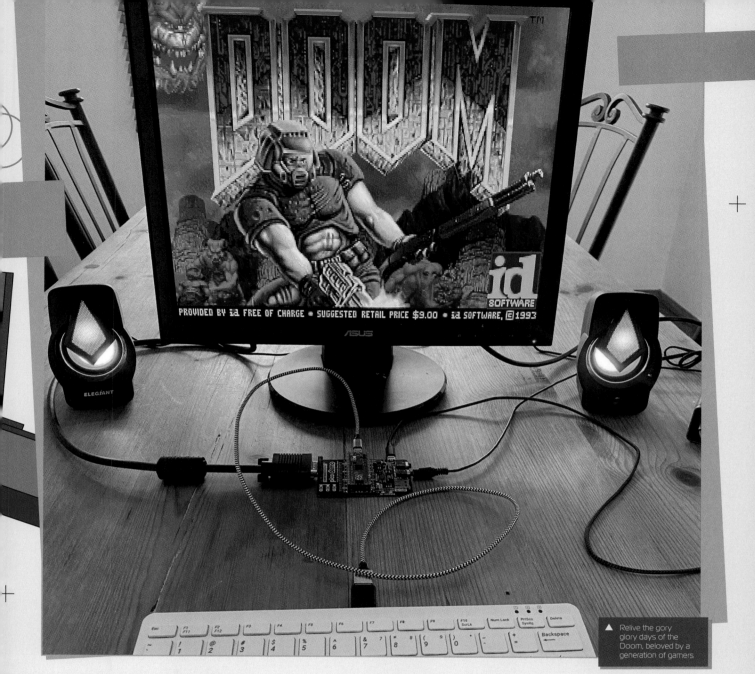

▲ Relive the gory glory days of the Doom, beloved by a generation of gamers

If you'd rather work with fresh hardware, we have the YouMakeTech's Pico 'GameBoy', a colour handheld console that wears its inspiration on its sleeve but is, in fact, designed to run and inspire the development of brand new games in MicroPython or C.

This little microcontroller can even play Doom, thanks to the gruelling porting and optimisation work invested by Raspberry Pi's in-house Pico SDK lead and classic games enthusiast Graham Sanderson, who's also responsible for the Pico's BBC Micro emulator (b-em) port.

Because this year marks the 50th anniversary of the first arcade release of Atari's Pong, on 29 November 1972, we've included the minimalist table tennis classic, showcasing a project that'll have you create a rainbow-colour handheld incarnation of the game, and demonstrating how to host a game of Pong on your local network by using Raspberry Pi Pico as a tiny web server. We're sure Pico W's future holds even more ambitious networked gaming applications.

> ❝ We're sure Pico W's future holds even more ambitious networked gaming applications ❞

Join us on a whirlwind tour of some of the coolest Pico gaming projects. Whether you're interested in hacking hardware, developing software, or just hooking up fully functional gaming experiences on the most cost-effective hardware around, there's something here to Pico your interest.

Play Doom
on Pico

We talk to Raspberry Pi Pico SDK lead Graham Sanderson about accurately porting Doom in all its glory to RP2040

Graham Sanderson, Raspberry Pi Pico SDK lead and performance architect by day, took on the challenge of making Doom run on the microcontroller in 2021. The game would be released in March 2022 after six months of development, but the first public evidence of the scheme appeared a year earlier, when Sanderson tweeted: "RP2040 Doom must be a thing, but if I do it, it needs to run all the demo WAD properly."

Sanderson grew up with the Sinclair ZX81 and BBC Micro. He says that the most fun

Alert!
Copyright

Video game files are protected by copyright law. Be sure to use ROM files that have been released with the owner's blessing, or modern homebrew games designed to be shared. There are lots of legal options.

magpi.cc/legalroms

▶ Input and output connections are neatly handled by the Pimoroni Pico VGA Demo Base

part of his role working on the RP2040 SDK has perhaps been, "taking 30-plus years of development and applying it to the constrained microcontroller environment, which has a similar feeling set of constraints to the home computers I had as a child."

He based RP2040 Doom on the Chocolate Doom (**chocolate-doom.org**) port of the original source code, itself designed to remove some of the limitations associated with the original DOS version.

The greatest challenge, he says, was getting the entire 4MB demo to fit, along with the game itself, into the Pico's 2MB flash storage. "The original game data file from Doom (DOOM1.WAD) is over 4MB big, and I have included everything: all the graphics, levels, music, sound, splash screens, multiplayer networking – the lot."

Particularly impressive is a tool called whd_ gen, part of the RP2040 Doom codebase, that converts and compresses WAD files into a custom format that reduces the level files' size by up to 57% in the case of DOOM1.WAD. WHD, which stands for 'Where's Half the Data?', is particularly effective at reducing Doom's graphical overheads, detailed in Sanderson's behind-the-scenes article on the techniques used to create this port.

Pico porting projects

Doom is by no means Sanderson's first venture into porting classic games to the RP2040: "The very first thing I ported was actually a ZX Spectrum emulator. It is on my list to open-source; it is just in a bit of a mess. This one was actually developed while we were still developing the RP2040 on FPGA and, so, was limited to a

system clock of 48MHz. It also, at one point, ran with less memory."

The ZX Spectrum emulator wasn't cycle accurate, but this didn't really matter as far as the performance of most Spectrum games is concerned. But that doesn't apply to every retro emulation system.

"Given the new-found freedom of a higher clock rate," – the production version of Raspberry Pi Pico has a maximum clock speed of 133MHz – "I thought I'd have a go at porting a BBC emulator where it is critical that everything happens at exactly the right cycle."

Sanderson says he does it for the challenge, to see if it's possible. That's a sentiment you'll hear from a lot of the coders and hardware hackers we've spoken to for this feature.

"I usually start with another code base, and of course they are never designed to run on something this constrained. [The] first thing is generally to divorce them from the

> ## ❝ He does it for the challenge ❞

idea that look-up tables many hundreds of kB big are a good idea, and [I] generally have to rewrite large portions with some sort of new methodology [or] approach and redo certain bits in assembly language."

RP2040 Doom shows the huge potential of one small microcontroller, a great deal of patience and determination, and three decades' worth of experience.

But you don't need to start with that level of experience to begin working on your own Raspberry Pi Pico projects. And if you prefer using other people's work as a platform for your own, higher-level projects, or as the basis of a really cool make, there's a wealth of kits and code to enjoy playing with.

Download your own Doom

You can download the release binary of RP2040 Doom at **magpi.cc/rp2040doom**, enter BOOTSEL mode, and copy the UF2 file you want over to it. To run it, Pico will have to be connected to a vgaboard-compliant RP2040 graphics board. The larger, 4MB file is designed for other RP2040-based devices, with more flash storage.

It's designed to run with a Pimoroni Pico VGA Demo Base (**magpi.cc/vgademobase**), which is

▲ RP2040 Doom has been designed to be faithful to the DOS original in graphics, sound, and responsiveness

functionally identical to the "VGA, SD card, and audio demo board" described in Raspberry Pi's 'Hardware Design with RP2040' documentation (**magpi.cc/rp2040hardware**). Sanderson notes, "It is a lot easier to use the VGA Demo Base, though!"

Although that's the easiest way to hook up display, controls, and sound, it's not the only way. In some of Sanderson's YouTube videos of the project, you'll see a breadboarded version of the setup, with the video GPIOs hooked up to a VGA connector via a resistor DAC and the audio ones to an PCM5102 I2S DAC board.

You can read a full account of Graham Sanderson's feat of optimisation at **magpi.cc/makingrp2040doom**.

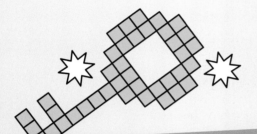

Hardware showcase:
RetroVGA

A fully-fledged retro computer in your pocket, with DB9 joystick input and VGA output

One of the most impressive Pico project kits around is Peter 'Bobricius' Misenko's RetroVGA, a Raspberry Pi Pico multi-retro computer which equips the microcontroller with a VGA output, DB9 joystick port, QWERTY keyboard, 3.5 mm audio output, integrated piezo speaker, USB port for input devices on supported emulators, and an SD card slot for storage.

The whole thing measures just 10 cm square, small enough to fit in a coat pocket, just in case you need to carry a retro microcomputer around with you. If you use a DB9 joystick, you'll need to reduce its cable length to around 30 cm, or it'll register false key presses.

The keyboard isn't quite full, lacking F-keys that are needed in some C64 titles, for example. The RetroVGA's sister device, the PICOZX (**magpi.cc/picozx128**) changes the number of GPIO pins devoted to VGA output to add extra keys, if that's a deal-breaker for you, but with a limited number of GPIO connections, compromises have to be made somewhere.

Once you've selected your game from the RetroVGA's SD card storage, the keyboard means you can LOAD and RUN it, and there are even directional keys to play with.

RetroVGA was first built to work with MCUME, Jean-Marc Harvengt's Multi CompUter Machine Emulator, and Miroslav Nemecek's PicoVGA display library. It now also supports Phil Scull's more feature-packed pico-zxspectrum emulator.

PicoVGA graphics standard
Harvengt and Nemecek were closely involved in the development. The RetroVGA board's video output wiring is compliant with the PicoVGA standard, and not with the Pico SDK's

▲ You can buy RetroVGA as a single PCB to mount your components on, as a pair with an optional top panel, or in its final form as a fully assembled multi retro computer

▲ With on-board controls, sound, and storage, RetroVGA is a compact colour for playing the latest games for the oldest of computers

supporting quicksave states, as well as USB keyboards and joypads.

Even when using a multi-device emulator such as MCUME, you'll use one emulator at a time. MCUME binaries are available to download at **magpi.cc/picoretrovga** and the UF2 files for pico-zxspectrum all live at **magpi.cc/zxuf2**.

While some of the projects in this overview of Pico-based retro gaming are still in their experimental stages, calling for breadboard assemblages or requiring you to have PCBs custom-made, you can just buy a RetroVGA kit at **magpi.cc/retrovga**.

Starting at $10 for the bare main PCB, $30 for a PCB and gold keyboard-bearing top panel, up to $108 for a fully assembled, ready-to-flash unit with all components, there are plenty of options available, depending on your needs and on how much soldering you're prepared to do.

default vgaboard standard. It uses different GPIO to VGA connections, which are documented in the project repository at **magpi.cc/picovga**.

Full documentation for RetroVGA itself is available at **magpi.cc/retrovgadocs** to help you

> " There are plenty of options available, depending on your needs and on how much soldering you're prepared to do "

▼ For best input results, you'll need to reduce the length of your DB9 joystick's cable to 30 cm

both assemble RetroVGA and get it up and running with a selection of emulators or the PicoVGA library, and there's enough detail there to get developers up and running

Currently, you can use MCUME to emulate the ZX81, ZX Spectrum, Atari 800, C64, VIC20, Atari 2600, Odyssey/Videopac, Colecovision, and Atari 5200. MCUME provides an effective, but bare-bones emulation experience. It can read game files from the SD card, giving you access to the rich world of C64 and Spectrum homebrew and indie releases. The pico-zxspectrum emulator is even more capable,

Hardware hacking

Four retro gaming hardware projects to test out

PICOCART64

The PicoCart64 is a flash cart for the Nintendo N64: a device that can hold code to be read by the original console hardware, allowing you to load your own games and software onto an actual N64.

It can boot N64 homebrew, which makes it a compelling way of accessing the console's rich post-market game ecosystem on real hardware, given that the most popular N64 flash carts are both costly and often hard to come by.

Creator Konrad Beckmann says that he set out on the project to find out if it's possible to use Raspberry Pi Pico instead of the FPGAs (field-programmable gate arrays) used in most commercially available flash carts, already expensive hardware that's become even more rare in the face of global chip shortages.

The PicoCart64 Lite is the project's first functional prototype flash cart, capable of loading N64 games, test files, and homebrew. It requires a single Raspberry Pi Pico, plus a MOSFET and resistor, and costs less than $10 to make.

The project's GitHub repository includes component lists, PCB schematics and, helpfully for those new to having their own PCBs fabricated, a link to a comparatively inexpensive service that'll do just that.

You'll find both hardware schematics and software in the develop branch of the PicoCart repository, and active discussion among contributors and hardware hackers implementing the tech on the Dubious Technology Discord channel.

Beckmann says that the PicoCart64 is "lowering the barrier, making it possible for people to make their own games and run it on actual hardware." Find N64 homebrew titles and SDKs to help you make your own at **magpi.cc/n64homebrew**.

magpi.cc/picocart64

PICO PONG

Software engineer Pip Austin has combined Raspberry Pi Pico with Pimoroni's rainbow LED Unicorn board (**magpi.cc/picounicorn**), a multicoloured 16×7 matrix of RGB LEDs with four integrated buttons, to create an eye-catchingly luminous game of Pong.

She chose Pong because, as she says, it's "a classic game with simple rules, but is visually very powerful." This aesthetic aspect is something that the Pico Unicorn really brought to life. Her own favourite parts of the game are visually arresting: the trailing light behind the ball, and the scrolling text when you win.

Released in 2021, Pico Pong was Austin's first venture into Python programming, and her self-written tutorial (**magpi.cc/picoponggame**) does a great job of showing her workings and explaining why she made each decision in her code, making it an inspiring resource for programmers new to the language.

But if you just want to get this shining, rainbow-bright Pong interpretation into your hands, you can find her complete MicroPython program in her GitHub repo, ready to copy to Pico with an IDE such as Thonny. She's hoping to make more games on her Pico in the future: "I have started work on Rat On A Scooter – making use of the scrolling function!"

magpi.cc/ponggit

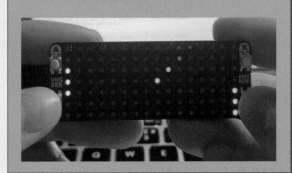

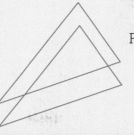

PICO-GB-CART

If you'd like to channel the spirit of Nintendo's classic handheld a bit more directly, you can always turn your Pico into a fully-fledged Game Boy flash cart. John Green was inspired by his experience making Game Boy emulators to see if he could make a hardware emulator with hardware skills he describes as "basic".

He used Gekkio's GB-BRK-CART (**magpi.cc/gbbrkcart**), a Game Boy breakout board, to interface with the handheld, learn what signals the Game Boy was expecting, and test to see if the Pico was fast enough to send I/O signals to it.

Then, he designed another prototype board that allowed for the Pico to be directly soldered to the PBC. It worked, but the way both these prototype boards interfaced with the Game Boy was a little unreliable, he says: "the Game Boy runs on 5 V and the Pico runs at 3.3 V, but after testing various voltage level converters, etc., I could never come up with a way of reading/writing data back and forth from the Game Boy and Pico whilst keeping the 'correct' voltages."

The make was fairly challenging: "You would think with the Game Boy only running at 4MHz and Pico being able to run at 133MHz base that [timing] should be no issue," Green says, but that didn't take into account time for the signal to be decoded by the Pico, get the next CPU instruction to send back to the Game Boy, and set the correct GPIO pins. In the end, Green found that 133MHz was too slow and Pico had to be overclocked to 360MHz. Only one of his ten Picos was able to function stably at this speed.

Green hasn't returned to a project's codebase for a while, but has plans for its future, including adding the ability to load game images from an SD card.

You can find instructions and software for the PICO-GB-CART at its GitHub repo, while schematics for the bespoke PCB created for the project by HDR can be found at **magpi.cc/ rp2040gbcart**.

Green's other Pico projects include a fighting game hitbox controller, and you can find an overview of personal software projects, such as a Game Boy raycaster, alongside a glimpse at his professional work in the games industry at **0xen.github.io**.

`magpi.cc/picogbcart`

YOUMAKETECH PICO GAMEBOY

One of the most aesthetically pleasing Raspberry Pi Pico projects we've seen is this Raspberry Pi Pico GameBoy, a complete playable console with a custom 3D-printed case, created by Vincent of YouMakeTech. Vincent supplies the STL files you'll need to create the case, a detailed parts list, connection diagrams, and a video assembly guide.

Although it's designed to look like Nintendo's classic handheld, albeit at three quarters the size of its original inspiration, the Raspberry Pi Pico GameBoy is no simple emulation console.

Equipped with a 1.54 in, 240×240, 65K colour screen, a D-pad, and two buttons, it exists to encourage tinkerers and would-be developers to explore game programming in their choice MicroPython or C++.

It's one of a series of Pico-based 3D-printed consoles that Vincent has created. If you're after a pure emulation console, YouMakeTech's Pico-GB (**magpi.cc/picogbemu**) can actually play DMG games designed for the original Game Boy (not the Game Boy Color) using the RP2040-GB emulator (**magpi.cc/rp2040gb**).

That includes the many modern homebrew Game Boy games that you can find on platforms such as Itch.io (**magpi.cc/gbgames**).

`magpi.cc/picogb`

Make
Pico Pong

Host HTML games, including Pong, on Pico W and play them over your local network

Most of the projects we've looked at in this feature will work perfectly with a standard Raspberry Pi Pico, but it's about time we did something that can take advantage of Pico W's unique networking capabilities. Let's host some web games.

There are limits to what you can host. Complex Twine games throw memory errors, for example. But as we're interested in retro games, we're in luck. To help celebrate Pong's 50th anniversary, here's a complete beginner's guide to hosting a two-player, HTML5 and JavaScript version of Pong on Pico, from installing MicroPython

Nuke it from orbit

If you start encountering memory errors, or need to get rid of files from previous projects or experiments, you can use the `os.remove` command to delete files one at a time from the REPL interpreter prompt. But if you need to comprehensively clear everything, MicroPython included, there's a more dramatic option.

Download **flash_nuke.uf2** from **magpi.cc/flashnuke** (direct file download). Unplug Pico (if connected), hold down the BOOTSEL button, and plug it in to your PC.

Copy **flash_nuke.uf2** over to it, and it'll reboot.

Don't forget to copy a fresh MicroPython nightly build over to it before embarking on your next project.

▼ A two-player game of Pong is served over your local network

01 Install Thonny
If you're using any Raspberry Pi computer to write to Pico, the Thonny Python IDE should already be installed. You'll find it in the repositories of other Linux distributions, while Windows and macOS users can download it from **thonny.org**.

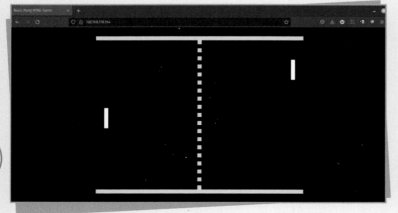

02 Install MicroPython
Grab a fresh nightly build of MicroPython from **magpi.cc/rp2picow**. Make sure Pico W is unplugged, hold down its BOOTSEL button, and simultaneously connect it to your PC. It'll appear as a mass storage device. Copy the UF2 file you downloaded over to it. Pico W will reboot.

03 Prepare Thonny
Open Thonny. In the bottom right of its window, you'll see a line of text indicating what

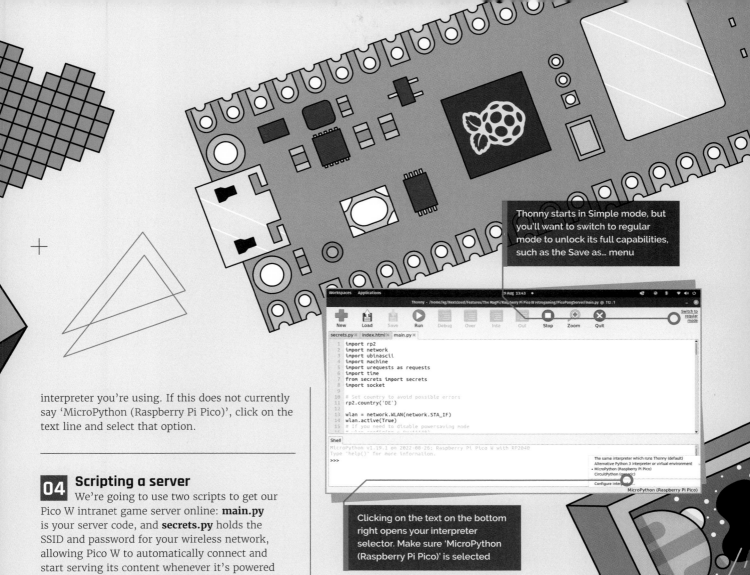

Thonny starts in Simple mode, but you'll want to switch to regular mode to unlock its full capabilities, such as the Save as... menu

Clicking on the text on the bottom right opens your interpreter selector. Make sure 'MicroPython (Raspberry Pi Pico)' is selected

interpreter you're using. If this does not currently say 'MicroPython (Raspberry Pi Pico)', click on the text line and select that option.

04 Scripting a server

We're going to use two scripts to get our Pico W intranet game server online: **main.py** is your server code, and **secrets.py** holds the SSID and password for your wireless network, allowing Pico W to automatically connect and start serving its content whenever it's powered on in vicinity of the network's access point. Our scripts are based on MIT-licensed project server scripts (**magpi.cc/picowledserver**) created by Nathan Bustler of Pi Cockpit. Download our

> 🔖 We're going to use two scripts to get our Pico W intranet game server online 🔖

versions of both scripts from our project page at **magpi.cc/picopongserver**.

05 Customise your network config

In Thonny, open our **main.py** and **secrets.py** scripts, update **secrets.py** with your own WiFi network's SSID and password, then save them to your connected Pico W. This is easiest if your copy of Thonny is in Regular mode rather than the Simple mode it starts in. If you're in Simple mode, click the 'Switch to regular mode' link at the top

right of Thonny's icon bar. In Regular mode, use the File menu's 'Save as...' option.

06 Just add Pong

We're going to use Straker's CC-licensed Basic Pong Game (**magpi.cc/basicpong**). Download our mirror of the HTML from **magpi.cc/github**, open it in Thonny and save the file to Pico. You can customise this, for example by changing the `ballSpeed` variable to make the ball move slowly.

07 Serve your balls!

In Thonny, select the **main.py** file that you saved to Pico, and press the play button. Assuming your networking has been correctly configured, you'll see lines displaying the device's MAC and IP addresses. Copy the IP address into the browser of a computer connected to the same local network. Congratulations, you and a friend (or your right and left hands) can now play a classic Pong clone in your browser. 🅼

Top Tip 👍

Further experiments

Try other lightweight HTML5 games and see how far you can push the Pico. It's also a great way to provide digital props and materials for physical games.

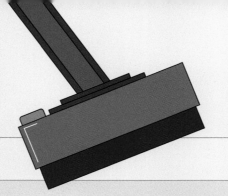

index.html

> Language: **HTML**

```html
001. <!DOCTYPE html>
002. <html>
003. <head>
004.   <title>Basic Pong HTML Game</title>
005.   <meta charset="UTF-8">
006.   <style>
007.   html, body {
008.     height: 100%;
009.     margin: 0;
010.   }
011.
012.   body {
013.     background: black;
014.     display: flex;
015.     align-items: center;
016.     justify-content: center;
017.   }
018.   </style>
019. </head>
020. <body>
021. <canvas width="750" height="585" id="game"></canvas>
022. <script>
023. const canvas = document.getElementById('game');
024. const context = canvas.getContext('2d');
025. const grid = 15;
026. const paddleHeight = grid * 5; // 80
027. const maxPaddleY = canvas.height - grid -
     paddleHeight;
028.
029. var paddleSpeed = 6;
030. var ballSpeed = 5;
031.
032. const leftPaddle = {
033.   // start in the middle of the game on the left
     side
034.   x: grid * 2,
035.   y: canvas.height / 2 - paddleHeight / 2,
036.   width: grid,
037.   height: paddleHeight,
038.
039.   // paddle velocity
040.   dy: 0
041. };
042. const rightPaddle = {
043.   // start in the middle of the game on the right
     side
044.   x: canvas.width - grid * 3,
045.   y: canvas.height / 2 - paddleHeight / 2,
046.   width: grid,
047.   height: paddleHeight,
048.
049.   // paddle velocity
050.   dy: 0
051. };
052. const ball = {
053.   // start in the middle of the game
054.   x: canvas.width / 2,
055.   y: canvas.height / 2,
056.   width: grid,
057.   height: grid,
058.
059.   // keep track of when need to reset the ball
     position
060.   resetting: false,
061.
062.   // ball velocity (start going to the top-right
     corner)
063.   dx: ballSpeed,
064.   dy: -ballSpeed
065. };
066.
067. // check for collision between two objects using
     axis-aligned bounding box (AABB)
068. // @see https://developer.mozilla.org/en-US/docs/
     Games/Techniques/2D_collision_detection
069. function collides(obj1, obj2) {
070.   return obj1.x < obj2.x + obj2.width &&
071.          obj1.x + obj1.width > obj2.x &&
072.          obj1.y < obj2.y + obj2.height &&
073.          obj1.y + obj1.height > obj2.y;
074. }
075.
076. // game loop
077. function loop() {
078.   requestAnimationFrame(loop);
079.   context.clearRect(0,0,canvas.width,canvas.height);
080.
081.   // move paddles by their velocity
082.   leftPaddle.y += leftPaddle.dy;
083.   rightPaddle.y += rightPaddle.dy;
084.
085.   // prevent paddles from going through walls
086.   if (leftPaddle.y < grid) {
087.     leftPaddle.y = grid;
088.   }
089.   else if (leftPaddle.y > maxPaddleY) {
090.     leftPaddle.y = maxPaddleY;
091.   }
092.
093.   if (rightPaddle.y < grid) {
094.     rightPaddle.y = grid;
095.   }
096.   else if (rightPaddle.y > maxPaddleY) {
097.     rightPaddle.y = maxPaddleY;
098.   }
099.
100.   // draw paddles
101.   context.fillStyle = 'white';
102.   context.fillRect(leftPaddle.x, leftPaddle.y,
     leftPaddle.width, leftPaddle.height);
```

```
103.    context.fillRect(rightPaddle.x, rightPaddle.y,
        rightPaddle.width, rightPaddle.height);
104.
105.    // move ball by its velocity
106.    ball.x += ball.dx;
107.    ball.y += ball.dy;
108.
109.    // prevent ball from going through walls by
        changing its velocity
110.    if (ball.y < grid) {
111.      ball.y = grid;
112.      ball.dy *= -1;
113.    }
114.    else if (ball.y + grid > canvas.height - grid) {
115.      ball.y = canvas.height - grid * 2;
116.      ball.dy *= -1;
117.    }
118.
119.    // reset ball if it goes past paddle (but only if
        we haven't already done so)
120.    if ( (ball.x < 0 || ball.x > canvas.width) &&
        !ball.resetting) {
121.      ball.resetting = true;
122.
123.      // give some time for the player to recover
        before launching the ball again
124.      setTimeout(() => {
125.        ball.resetting = false;
126.        ball.x = canvas.width / 2;
127.        ball.y = canvas.height / 2;
128.      }, 400);
129.    }
130.
131.    // check to see if ball collides with paddle. if
        they do change x velocity
132.    if (collides(ball, leftPaddle)) {
133.      ball.dx *= -1;
134.
135.      // move ball next to the paddle otherwise the
        collision will happen again
136.      // in the next frame
137.      ball.x = leftPaddle.x + leftPaddle.width;
138.    }
139.    else if (collides(ball, rightPaddle)) {
140.      ball.dx *= -1;
141.
142.      // move ball next to the paddle otherwise the
        collision will happen again
143.      // in the next frame
144.      ball.x = rightPaddle.x - ball.width;
145.    }
146.
147.    // draw ball
148.    context.fillRect(ball.x, ball.y, ball.width,
        ball.height);
149.
150.    // draw walls
151.    context.fillStyle = 'lightgrey';
152.    context.fillRect(0, 0, canvas.width, grid);
153.    context.fillRect(0, canvas.height - grid,
        canvas.width, canvas.height);
154.
155.    // draw dotted line down the middle
156.    for (let i = grid; i < canvas.height - grid; i +=
        grid * 2) {
157.      context.fillRect(canvas.width / 2 - grid / 2, i,
        grid, grid);
158.    }
159.  }
160.
161.  // listen to keyboard events to move the paddles
162.  document.addEventListener('keydown', function(e) {
163.
164.    // up arrow key
165.    if (e.which === 38) {
166.      rightPaddle.dy = -paddleSpeed;
167.    }
168.    // down arrow key
169.    else if (e.which === 40) {
170.      rightPaddle.dy = paddleSpeed;
171.    }
172.
173.    // w key
174.    if (e.which === 87) {
175.      leftPaddle.dy = -paddleSpeed;
176.    }
177.    // a key
178.    else if (e.which === 83) {
179.      leftPaddle.dy = paddleSpeed;
180.    }
181.  });
182.
183.  // listen to keyboard events to stop the paddle if
        key is released
184.  document.addEventListener('keyup', function(e) {
185.    if (e.which === 38 || e.which === 40) {
186.      rightPaddle.dy = 0;
187.    }
188.
189.    if (e.which === 83 || e.which === 87) {
190.      leftPaddle.dy = 0;
191.    }
192.  });
193.
194.  // start the game
195.  requestAnimationFrame(loop);
196.  </script>
197.  </body>
198.  </html>
```

main.py

> Language: **Python**

```python
001.  import rp2
002.  import network
003.  import ubinascii
004.  import machine
005.  import urequests as requests
006.  import time
007.  from secrets import secrets
008.  import socket
009.
010.  # Set country to avoid possible errors
011.  rp2.country('DE')
012.
013.  wlan = network.WLAN(network.STA_IF)
014.  wlan.active(True)
015.  # If you need to disable powersaving mode
016.  # wlan.config(pm = 0xa11140)
017.
018.  # See the MAC address in the wireless chip OTP
019.  mac = ubinascii.hexlify(
      network.WLAN().config('mac'),':').decode()
020.  print('mac = ' + mac)
021.
022.  # Other things to query
023.  # print(wlan.config('channel'))
024.  # print(wlan.config('essid'))
025.  # print(wlan.config('txpower'))
026.
027.  # Load login data from different file for safety
      reasons
028.  ssid = secrets['ssid']
029.  pw = secrets['pw']
030.
031.  wlan.connect(ssid, pw)
032.
033.  # Wait for connection with 10 second timeout
034.  timeout = 10
035.  while timeout > 0:
036.      if wlan.status() < 0 or wlan.status() >= 3:
037.          break
038.      timeout -= 1
039.      print('Waiting for connection...')
040.      time.sleep(1)
041.
042.  # Define blinking function for onboard LED to
      indicate error codes
043.  def blink_onboard_led(num_blinks):
044.      led = machine.Pin('LED', machine.Pin.OUT)
045.      for i in range(num_blinks):
046.          led.on()
047.          time.sleep(.2)
048.          led.off()
049.          time.sleep(.2)
050.
051.  # Handle connection error
052.  # Error meanings
053.  # 0  Link Down
054.  # 1  Link Join
055.  # 2  Link NoIp
056.  # 3  Link Up
057.  # -1 Link Fail
058.  # -2 Link NoNet
059.  # -3 Link BadAuth
060.
061.  wlan_status = wlan.status()
062.  blink_onboard_led(wlan_status)
063.
064.  if wlan_status != 3:
065.      raise RuntimeError('Wi-Fi connection failed')
066.  else:
067.      print('Connected')
068.      status = wlan.ifconfig()
069.      print('ip = ' + status[0])
070.
071.  # Function to load in html page
072.  def get_html(html_name):
073.      with open(html_name, 'r') as file:
```

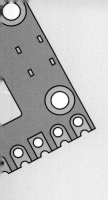

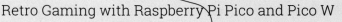
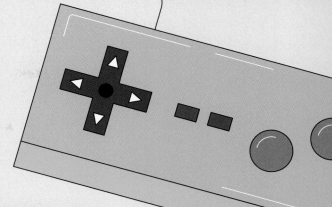

secrets.py

> Language: **Python**

```python
001. secrets = {
002.     'ssid': 'Enter_your_SSID_here',
003.     'pw': 'Enter_your_Wi-Fi_password_here',
004.     }
```

```python
074.         html = file.read()
075.
076.     return html
077.
078. # HTTP server with socket
079. addr = socket.getaddrinfo('0.0.0.0', 80)[0][-1]
080.
081. s = socket.socket()
082. s.bind(addr)
083. s.listen(1)
084.
085. print('Listening on', addr)
086. led = machine.Pin('LED', machine.Pin.OUT)
087.
088. # Listen for connections
089. while True:
090.     try:
091.         cl, addr = s.accept()
092.         print('Client connected from', addr)
093.         r = cl.recv(1024)
094.         # print(r)
095.
096.
097.         response = get_html('index.html')
098.         cl.send('HTTP/1.0 200 OK\r\nContent-type:
     text/html\r\n\r\n')
099.         cl.send(response)
100.         cl.close()
101.
102.     except OSError as e:
103.         cl.close()
104.         print('Connection closed')
105.
106. # Make GET request
107. #request = requests.get('http://www.google.com')
108. #print(request.content)
109. #request.close()
```

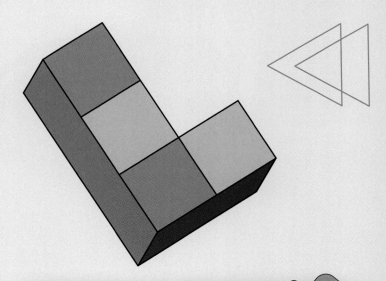

Reviews and Resources

176

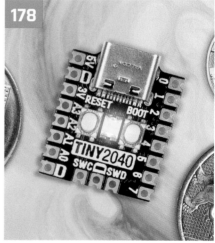

178

179

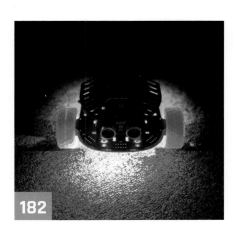

182

190

192

Argon **EON**

▶ Argon40 ▶ **argon40.com** ▶ From £95 / $128

SPECS

DIMENSIONS:
210 mm × 165 mm
(per side)

STORAGE
CAPACITY:
2 × 3.5˝ drives,
2 × 2.5˝ drives,
M.2 via internal
USB adapter

TRANSFER
SPEEDS:
HDD 7200 rpm:
150 MBps,
SSD 7200 rpm:
280 MBps, Max:
500 MBps

PERIPHERALS:
IR receiver, OLED
screen

Argon adds to its respected Raspberry Pi cases with a striking new NAS enclosure. **PJ Evans** takes it for a spin

Sleek lines, tailored styling, and impressive build quality hallmarked the original Argon40 case. Now Argon is back with a follow-up, the EON. Having raised over £100,000 on Kickstarter, the project is nearly ready to ship, and we've been lucky enough to road-test a prototype for a couple of months.

The EON is a full network-attached storage (NAS) enclosure for your Raspberry Pi 4. Whereas previous NAS projects often involve a USB hub and cables to external drives, the EON fits up to four drives in an enclosure with a single power source. This allows you to build your own multi-drive storage solution for a fraction of the price of a PC-based equivalent.

The first thing that strikes you about the EON is the enclosure itself. No boring black box here – instead we appear to have a pyramidical version of the monolith from *2001: A Space Odyssey*. The frame is built from "space-grade" aluminium, and feels reassuringly heavy. If dropped, we suspect it's the floor that's getting damaged, not the internals. Two magnetically-held opaque plates give a sleek black finish. The rear is solid with plenty of mounting holes for your drives. Raspberry Pi 4 sits at the base of the unit.

Inside is a layered affair with Raspberry Pi 4 at the bottom, then two further PCBs, the top exposing four SATA ports for your drives. These connect via USB 3.0 directly to your Raspberry Pi 4 using a supplied external connector. The SATA ports support four drives, mounted vertically. The unusual shape of the EON means that the two outermost connectors only support 2.5˝ drives, but the inner two can be full-size 3.5˝ HDD or SSD.

Versatile

Argon has also baked versatility into the design with a built-IR port that works with the firm's remote control (reviewed last month), perfect for a home theatre setup. A fan provides cooling for the entire unit, and the top-mounted power switch is a small OLED screen that can be customised to display anything, such as IP address or CPU temperature.

Installing drives was straightforward and the supplied screws and tools were a welcome bonus. Our Raspberry Pi 4 immediately recognised the new devices, and soon we had a RAID system up and running. Argon plans to add an internal USB port to support an M.2 SSD, so you can separate the operating system and RAID drives.

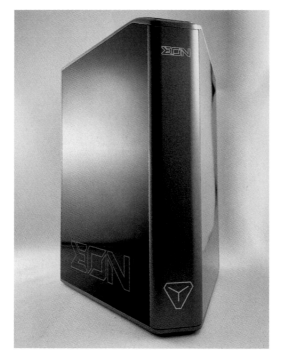

▶ Stunning looks, solid build, and lots of space for storage

We found it hard to come up with anything negative about this case. The positioning of the power switch makes it very easy to hit it when handling the device, which can punish the impatient when moving things around. Also, despite the rock star looks, the shape of the device does cheat you out of 4 × 3.5″ drives, which could be seen as form over function.

> ❝ We appear to have a pyramidical version of the monolith from 2001: A Space Odyssey ❞

These minor nitpicks aside, this is one of the most impressive home NAS setups we've seen. If you want your home server not only capable of huge storage and Raspberry Pi performance, but also as something great to look at, this is for you. Ⓜ

▲ Shown here with a single 2.5″ SSD drive. Note the adaptive cooling fan at the top

▲ All of Raspberry Pi's ports are available, including full-size HDMI, with the SATA connectors sitting above

Verdict

This is simply a stunning piece of engineering. You can build a NAS that competes with much more expensive options, and it looks like something from a sci-fi movie.

9/10

Tiny **2040**

▶ Pimoroni ▶ **magpi.cc/tiny2040** ▶ From £7 / $7

A truly tiny RP2040-powered microcontroller board. By **Phil King**

Raspberry Pi Pico is only the size of a stick of gum, but if you need an even smaller microcontroller, the Tiny 2040 could fit the bill. Roughly the dimensions of a regular UK postage stamp, this little board is about 40% the length of a Pico, but packs the same powerful RP2040 system-on-chip.

> **The Tiny 2040 adds a very welcome on-board reset button, which saves repeatedly unplugging it**

You can flash it with the Pico firmware and program it with MicroPython, C/C++, or CircuitPython. So it's very versatile and easy to get started with.

To achieve a smaller footprint, a few compromises have been made. Most notably, the Tiny 2040 only has a total of 16 pins versus Pico's 40, plus a debug header. Of these, there are 12 GPIO pins, compared to 26 on Pico. So it won't be compatible with most Pico add-on modules. As with Pico, there's no wireless connectivity.

On the plus side, the Tiny 2040 does break out a fourth ADC input – which is connected to an on-board temperature sensor on Pico.

Tiny choices

You can buy the Tiny 2040 with or without (2.54mm pitch) header pins attached. There's also a choice between 2MB of on-board QSPI flash storage (as on Pico) or 8MB. The Tiny 2040 is a fair bit pricier than Pico itself, which is only £3.60 / $4.

On the plus side, the Tiny 2040 adds a very welcome on-board reset button, which saves repeatedly unplugging and reconnecting the board. There's also a programmable RGB LED (linked to three internal GPIO channels), rather than the single-colour green one on Pico.

Other than that, its advantages are mainly down to its diminutive size, making it suitable for wearable projects or even a tiny robot (e.g. **magpi.cc/tiny2040robot**). While not quite as minuscule as a Nionics Atto or Seeeduino XIAO, its RP2040 chip is a lot more powerful. M

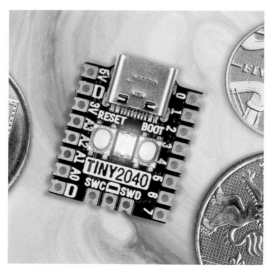

▶ "I can do anything you can do smaller!" The Tiny 2040 is a fraction of the size of a Pico

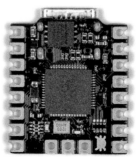

▲ The RP2040 SoC is located on the underside, which may make the board a little trickier to surface-mount

Verdict

Perfect for wearables and other portable projects, it packs the same RP2040 SoC as Pico but on a much smaller board, plus a few bonus features.

8/10

Badger **2040**

▶ Pimoroni ▶ **magpi.cc/badger2040** ▶ £12 / $16

SPECS

SCREEN:
2.9-inch black and white e-ink display (296 × 128 pixels)

PROCESSOR:
RP2040 (dual Arm Cortex-M0+ running at up to 133MHz with 264kB of SRAM); 2MB of QSPI flash supporting XiP

I/O:
Five front user buttons; Reset and boot buttons; White LED USB-C connector for power and programming; JST-PH connector for attaching a battery (input range 2.7V – 6V)

Raspberry Pi 2040-based e-ink badge and buttons make for an interesting device. By **Lucy Hattersley**

R aspberry Pi 2040 is the chip at the heart of Raspberry Pi Pico, and it's making its way across the tech ecosystem, powering a range of unique devices.

Badger 2040 is an RP2040 mounted on a 2.9-inch black and white e-ink display with five buttons.

The device comes fully assembled, so all you need to do is turn it on and load up the latest version of the software from Pimoroni's GitHub page (**magpi. cc/pimoronipicoreleases**). Installation is a case of holding down the BOOT button to mount RP2040 onto the destkop, and copying the UF2 file to the mounted storage.

There's an app

There is a clock app, ebook app (pre-loaded with *Wind in the Willows*), an image app, interactive list app, badge app, along with a QR code, info, and help information displays.

As with many projects, the fun begins when you start exploring what you can do with Badger 2040 in a coding environment. Pimoroni's documentation is, as typical, excellent, including a Getting Started with Badger 2040 guide (**magpi.cc/getstartedbadger**).

▲ On the rear of Badger 2040 sits the RP2040 chip along with BOOT and RST buttons and a white LED

" An optional accessory kit includes a AAA battery holder "

The tutorials take you through writing 'Hello Badger' to the screen and customising the default apps. This is done by exploring a range of text files in Thonny (using View > Files). Images can be converted using Pillow and the **convert.py** file

An optional accessory kit includes a AAA battery holder and batteries, Velcro square, lanyard, and cable. You can also power Badger 2040 by connecting a lithium battery to the JST-PH connector.

Moving on from the default Badger OS and test projects, Badger 2040 has a Qwiic/STEMMA QT port for connecting breakouts with a JST-SH cable and STEMMA QT adapter (**magpi.cc/stemmaqt**). With this, you can explore integration with a variety of sensors, breakout boards, and accessories. ◼

Verdict

Badger 2040 is a fun accessory that integrates nicely with Raspberry Pi thanks to its RP2040 base. Simple to set up, but a lot of potential for integration with your projects.

8/10

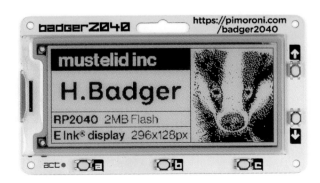

▲ Badger 2040 features an e-ink display and five navigation buttons

▲ The rear features the battery connector, edge connector, and power and boot /USR buttons

Tufty **2040**

▶ Pimoroni ▶ **magpi.cc/tufty2040** ▶ From £24 / $24

With a colour LCD screen, is this the ultimate interactive name badge? **By Phil King**

SPECS

DISPLAY:
2.4 in colour IPS LCD display, 320×240 pixels

POWER:
JST-PH battery connector (input range 3–5.5 V), USB-C

FEATURES:
5 × user buttons, LED, Qwiic/STEMMA QT port, breakout edge connector (I2C, UART, SWD), 8MB flash storage

The Tufty 2040 badge is based on Raspberry Pi's RP2040 microcontroller chip, as used on Pico. It goes one better than the Badger 2040 (reviewed in issue 116, **magpi.cc/116**) with the use of a 2.4-inch colour LCD screen in place of a monochrome e-ink display.

With a rapid refresh rate, the LCD enables the Tufty to do a lot more, including showing animated text and graphics. A simple game is even included as one of the preloaded examples, selectable via a menu when you turn it on. Others include wavy scrolling text, Pride and retro badge layouts, and an old-school 'Sketchy-Sketch' drawing tool. As on the Badger, control is via five programmable user buttons.

Connecting the Tufty 2040 to a computer via USB enables you to program it in MicroPython or C++. The PicoGraphics library makes it relatively easy to write text and draw shapes. JPEGs can also be rendered, and sprites imported from a sprite sheet.

Wear it well

For portable use, a JST-PH battery connector accepts input from 3V to 5.5V. An optional Accessory Kit includes a 3×AAA battery pack, Velcro pad to fix it to the rear, and lanyard. A less bulky alternative is to use a LiPo battery, although

the Tufty has no circuitry to charge it. Naturally, power drain is higher than using an e-ink display: around 100 mA in total. The on-board light sensor could be used to auto-dim the display via PWM, however.

A neat bonus feature is the Qwiic/STEMMA QT port which can be used to connect I2C sensors and other add-ons. So you could even use the Tufty 2040 as a data display instead of a badge. **M**

> ❝ With a rapid refresh rate, the LCD enables the Tufty to do a lot more ❞

▲ The Tufty 2040 displaying the Pride badge example on its colour LCD

Verdict

The crisp colour screen makes for a super-cool interactive name badge that is versatile and fairly easy to program.

9/10

Pico Unicorn Pack

▶ Pimoroni ▶ **magpi.cc/picounicorn** ▶ £22 / $23

SPECS

FEATURES:
16×7 RGB LED matrix, 4 × programmable push-buttons

DIMENSIONS:
65 × 25 × 10 mm

Light up your Pico projects with this vibrant LED matrix. By **Phil King**

Want to add some sparkle to your Pico project? The Pico Unicorn Pack offers a 16×7 matrix of bright RGB LEDs – that's 112 in total – along with four programmable push-buttons. All you need is a Raspberry Pi Pico with soldered male pins and you can plug it into the Pico Unicorn's dual female headers, found on the rear of the board.

You'll also need to flash Pico with Pimoroni's custom MicroPython UF2 firmware. This

> ## 🙵 The LEDs are updated in the background with very little CPU usage 🙶

includes the MicroPython module for the Pico Unicorn, which you can import at the top of your programs. While not as sophisticated as the Python one for Pimoroni's Unicorn HATs for Raspberry Pi computers, it does enable you to set individual pixels to RGB values (or a single value) and read button presses. Disappointingly, there's no built-in function for scrolling text, although it can be done with a bit of know-how.

▲ The same size as a Pico, the Unicorn has an RGB LED matrix and four push-buttons

Pico Pong

Only two MicroPython code examples for Pico Unicorn are provided in Pimoroni's GitHub repo, but doing a search for 'Pico Unicorn' on GitHub reveals a wide variety of programs created by the community, giving you an idea of the possibilities. Our favourite is a two-player game of Pong using the buttons as controls (**magpi. cc/unicornpong**). There's even a funky plasma generator that enables you to scroll messages using a frame buffer (**magpi.cc/unicornplasma**). We also found a version of Conway's Game of Life on the Pimoroni forums.

Those RGB LEDs are bright and very responsive. Controlled using the PIO state machines on Pico's RP2040 chip, they're updated in the background with very little CPU usage. In fact, it's so fast that 14 bits of resolution can be achieved, resulting in smoother brightness transitions using gamma-corrected values. In short, it looks very impressive. 🄼

▼ The RGB LEDs are bright and super-responsive, making use of Pico's PIO

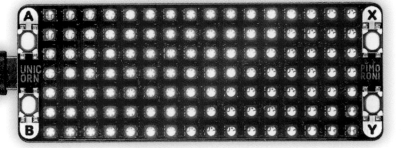

Verdict

With a matrix of super-responsive RGB LEDs, it's ideal for animations and even games using the tiny push-buttons.

9 /10

Trilobot

SPECS

FEATURES:
2 × FR-4 PCBs, mounts for ultrasonic distance sensor (supplied) and camera, six-zone RGB underlighting, 4 × push-buttons with status LEDs

MOTORS:
110:1 metal-gear motors with pre-soldered shims; DRV8833PWP motor controller integrated into PCB

EXPANSION:
2 × Qwiic / STEMMA QT ports, 1 × servo header, 1 × I2C header, 5 × Breakout Garden headers (all unpopulated)

▶ Pimoroni   ▶ **magpi.cc/trilobot**   ▶ From £48 / $54

A two-wheeled robot with style and plentiful options for expansion. By **Phil King**

At first sight, it may look like most other two-wheeled Raspberry Pi robots, but Trilobot has a lot of neat tricks up its sleeve, including some very cool LED underlighting. Described by Pimoroni as a 'mid-level' robot, it's designed to be simple for newcomers to get started with, while offering plenty of possibilities for adding extra functionality via its ports and numerous unpopulated headers.

The standard Base Kit includes everything you need, apart from a Raspberry Pi 4, microSD card, USB-C power bank, and optional Raspberry Pi Camera Module v2. Note, however, that you could use any full-size Raspberry Pi model and/or power bank by substituting the supplied USB-C cable with one for the connections you need.

Easy assembly

We found the kit very straightforward to assemble, aided by the well-illustrated step-by-step online guide. The only real difficulty we encountered was getting the tiny nuts into the plastic brackets for

the motors – the fit is tight, so we needed to use the blade of a flat-headed screwdriver to push them into place.

The robot's chassis comprises two FR-4 PCBs. Rather neatly, a DRV8833PWP dual H-bridge motor controller is integrated seamlessly into the main PCB, with two mini JST sockets to connect the supplied short cables to the pre-soldered shims of the metal-gear motors – no soldering or screwing required. Two moon buggy wheels are supplied, along with a standard metal ball castor to be attached to the rear of the board (with an acrylic spacer).

At this point, you can add an optional Raspberry Pi Camera Module v2 using the kit's camera mount before adding the supplied HC-SR04 ultrasonic distance sensor in its own mount in front, with a hole for the camera lens to poke through as the two are sandwiched together. The distance sensor's pins fit into a port on the PCB.

▲ Some spectacular underglow effects are possible using the six RGB LEDs on the underside of the main PCB

▶ The supplied ultrasonic distance sensor and optional Camera Module are mounted at the front of the robot

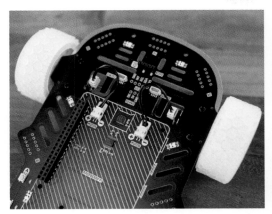

▲ Wiring up motors has never been easier: just plug each cable into the mini JST connectors on the motor and PCB port

▲ The components supplied in the Trilobot Base Kit. Just add a Raspberry Pi, microSD card, power bank, and Camera Module (optional)

Your Raspberry Pi is then mounted upside down on the main PCB's female GPIO header (via a booster). Once attached using the standoffs, the top PCB serves as a platform for the power bank – ours was a fair bit wider than the one featured in Pimoroni's assembly guide, but the Velcro straps secured it well enough.

Robot running

With the robot assembled, you just need to install the software in Raspberry Pi OS using three Terminal commands. A comprehensive Python library offers a large range of useful functions and comes with several code examples to help you get started. These include some impressive underlighting effects using the six RGB LEDs located on the bottom of the main PCB, reading the four push-buttons located near the rear, and avoiding walls using the distance sensor.

Another code example enables you to remote-control the Trilobot wirelessly using a Bluetooth Xbox, PlayStation, or 8BitDo gamepad. There aren't yet any examples for using the camera, although it should be fairly straightforward to use it to take photos and stream video; you could even use OpenCV to add object or face detection.

▲ One of the unpopulated Breakout Garden headers is right at the front, although we're not quite sure how practical it is as it's so near to the ultrasonic distance sensor

> " Some impressive underlighting effects using the six RGB LEDs located on the bottom of the main PCB "

While it's not the speediest robot we've ever seen, the 110:1 metal-gear motors provide a good level of torque while the moon buggy wheels offer plenty of grip, even on a hard surface.

There are plentiful possibilities for expanding the robot. As well as two Qwiic / STEMMA QT ports, the main PCB has several unpopulated headers: one for a 5V servo (or NeoPixel strips if you prefer), another for I2C, along with five more for adding Pimoroni's Breakout Garden sockets to use the wide range of sensors available. So, for instance, you could add a motion sensor or mini LCD. 🔟

Verdict

With a solid chassis, detailed software library, and abundant options for expansion, it all adds up to an excellent robotics platform for beginners and more experienced enthusiasts alike.

9/10

Autonomous Robotics Platform
for Raspberry Pi Pico

SPECS

DIMENSIONS:
PCB length:
126 mm; PCB
width: 80 mm;
wheel diameter
(with tyre):
67.5 mm

SENSORS:
Ultrasonic
distance sensor
HC-SR04 5 V;
Kitronik line-
following
sensor board

MOTORS:
2 × TT
geared motors

▶ Kitronik ▶ **magpi.cc/kitronikpico** ▶ £41 / $49

This fun robotics board uses Pico to great effect.
Lucy Hattersley takes it for a spin

Kitronik's Autonomous Robotics Platform caught our attention recently thanks to its usage of Raspberry Pi Pico, rather than the more common Raspberry Pi Model B or Zero models.

The kit contains a robotics platform chassis with two TT motors pre-mounted. Two large yellow wheels are attached to the side, along with an ultrasonic sensor on the top and a line-following sensor underneath. Finally, a Pico or Pico W with a GPIO header soldered in can be mounted in the middle of the two motors. Four AA batteries are slotted in underneath to provide power to the motors and Pico.

The result is a cute-looking robot that is easy to assemble, making it perfect for younger robotics makers. It's incredibly lightweight and moves around at a brisk pace.

All aboard

There are a few extras on the robot chassis worth mentioning. On all four corners sit ZIP LEDs that add bling (and can be useful for feedback). A hole in the middle of the board is used to hold a marker pen for turtle-like drawing. There is an on/off switch to cut the power and a button that responds to code (as opposed to the BOOTSEL button on Pico W). Finally, there's an on-board buzzer to make audio feedback.

We found it easy to set up, thanks to the included manual. At least to the point where the physical assembly was complete. Following the build, the manual skims over the API and mostly directs you to the Kitronik website (**kitronik.co.uk/pico-arp-motors**) for more detail on how to code and control the robot.

Clone the corresponding GitHub repo (**magpi.cc/kitpicogit**), and you'll discover code to go with all the tutorials and some great example programs. Along with tests for all the motors, sensors, button, and buzzer, there's code that runs the robot around in circles, line-following examples, pen-lifting examples, and a program that uses the sensors to control the lights.

▶ Four LEDs light up the Kitronik Autonomous Robotics Platform

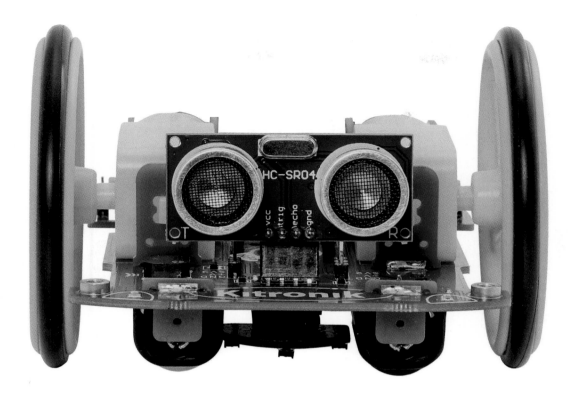

Autonomous Robotics Platform is a good-looking robot that's easy to control

The GitHub page has documentation on the API, and the tutorials are comprehensive.

Using Pico instead of Raspberry Pi for the code has advantages and disadvantages. Even though Pico W is now available, you cannot remote-control the robot via a web or smartphone app (as you can with many other robots). Perhaps this functionality can be implemented down the line.

❝ Ideal for a cost-minded learning environment ❞

Pico runs code as soon as it's switched on, though, so the robot is functional in a code-and-drop way that makes it more reliable than Raspberry Pi running a full OS. And you're not faced with the usual SSH and wireless networking complication that troubles many a robot setup. You create code on your computer and drop it directly onto Pico to run.

We think this is a nice robot build that will be lots of fun. It packs a lot of features onto a board given its low cost (which is better value when you factor in running Pico, rather than a full-blown Raspberry Pi computer).

Autonomous Robotics Platform is ideal for a cost-minded learning environment.

These robots are cheap to buy, easy to set up, sturdy, and fun to program. 🅜

The PCB board comes with two TT motors mounted; wheels, sensors, and Pico are attached to complete the build

Verdict

A great little robotics learning environment that is great value when you factor in the low cost of Pico. It's packed with features too.

9/10

Weather HAT +
Weather Sensors Kit

▶ Pimoroni ▶ **magpi.cc/weatherhat** ▶ £83 / $94

Come rain or shine, you'll know all the details with this fully-fledged weather station. By **Phil King**

SPECS

SENSORS:
Wind speed, wind direction; BME280 rainfall, temperature, pressure, humidity; LTR-559 light sensor

FEATURES:
1.54-inch colour LCD, 4 × push-buttons, 2 × RJ11 sockets, on-board Nuvaton microcontroller with 12-bit ADC

EXPANSION:
4 × analogue inputs, I2C header

We've seen various Raspberry Pi DIY weather station projects and commercial kits before, but the Weather HAT makes setup a whole lot easier.

For starters, the board features standard telecoms-style RJ11 ports that enable you to just plug in the connectors from the wind speed, wind direction (vane), and rainfall sensors. Similar to the ones formerly available from Maplin stores, these three plastic sensors are provided in the full Weather HAT kit – or, if you already have some, you can buy the Weather HAT on its own for £36 / $41.

Secondly, the Weather HAT has an on-board Nuvoton microcontroller with a 12-bit ADC to read analogue signals from the sensors reliably. A nice bonus is that four extra analogue input channels are broken out in an unpopulated header on the bottom edge of the board (along with 3V and GND), so you could add extra sensors. Not only that, but there's an I2C header on the underside of the board.

Whether you'll need extra sensors is a moot point: the HAT already incorporates a standard BME280 temperature, pressure, humidity sensor and an LTR-559 light sensor – as featured on Pimoroni's earlier Enviro boards. A 1.54-inch colour LCD screen completes the package, with four tiny push-buttons around it.

Station setup

Assembling the weather station takes only around 15 to 20 minutes. The wind direction and speed sensors are screwed onto either end of a sturdy plastic arm that fits into the top of a two-section hollow metal mast. A shorter arm is clamped to the mast and holds the rain sensor. We then planted the metal mast into some garden soil, but clamps

▶ The rear of the Weather HAT. It will fit any 40-pin Raspberry Pi model and also features an I2C header for extra inputs

▲ It's very easy to hook your weather station up to an Adafruit IO web dashboard to log all the data there

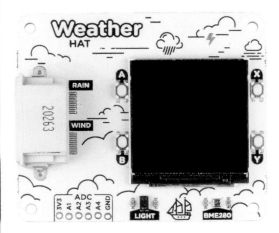

are also supplied in case you want to secure it to a drainpipe or something similar.

As mentioned, the sensors' RJ11 connectors simply plug into the labelled sockets on the Weather HAT. If you're wondering why there are only two sockets, this is because the wind speed sensor's connector fits into a socket on the wind vane, and then wires from both are routed to one RJ11 connector into the HAT.

The cables are approximately 3 m long, so unless you use extenders, your Raspberry Pi needs to be located fairly close by. Naturally, if this is outside, you should place it in a weatherproof box (not supplied). You may also need a mains power extension. Alternatively, you could place

> ## The sensors' RJ11 connectors simply plug into the labelled sockets on the Weather HAT

your Raspberry Pi indoors, although you then wouldn't get accurate outdoor temperature and humidity readings.

Software library
As usual from Pimoroni, there's a fully-fledged Python library for the Weather HAT, complete with code examples. The main **weather.py** example shows readings from the sensors on the mini

▲ The **weather.py** Python code example shows live data from the sensors on the LCD

▶ The Weather Sensors Kit comprises wind speed, wind direction, and rain sensors, mounted on a metal mast

LCD, with push-buttons used to switch between individual sensor readings and between a digital readout or bar graph. We particularly liked the wind display, with an arrow rotating to show the direction, its size indicating the speed. Note that you may well need to use a compass and rotate the metal mast so that the wind vane is oriented to obtain accurate direction readings.

In the long term, you will want to keep a log of your data and view it on a web dashboard, which is where the Adafruit IO code example comes in. Just sign up for a free account at **adafruit.com** and you can build your own custom dashboard from preset blocks such as line graphs and gauges, linking them to the already created feeds coming from the Weather HAT. Within minutes we had our own weather web dashboard set up.

▲ Along with on-board sensors and a mini LCD to show data, the Weather HAT has inputs for wind and rain sensors, plus a breakout header

Verdict

Packed with all the features and expandability you could need, this comprehensive kit is very good value for money and makes it a breeze to set up your own weather station.

9 /10

PecanPi **Streamer**

▶ Orchard Audio ▶ **magpi.cc/streamer1** ▶ £480/$550

This black box contains some of the highest quality audio components available alongside a Raspberry Pi 3B, but is it worth it? **PJ Evans** engages his golden ears.

Raspberry Pi continues to make waves in the world of high-end audio. At the very top of the pile is Orchard Audio. This company cares not for flashing lights, gimmicks or even Bluetooth, but instead is dedicated to one thing and one thing only: producing the best sound possible. Orchard's unique proposition is the use of the highest quality components, right down to the resistors. So it was with great anticipation that we received the PecanPi Streamer v3.

Minimalist

This case is minimalism taken to a new level. Looking at the rear, we see the ports from a Raspberry Pi 3B, and not only the usual phono (RCA) sockets but XLR sockets too. This unassuming box is aimed right at the professional market as well as the audiophile.

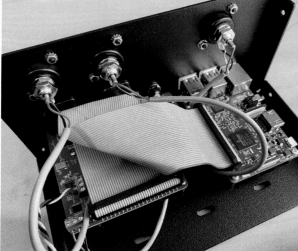
▲ A familiar Raspberry Pi 3B sits next to state-of-the-art audio components

> **❝ This case is minimalism taken to a new level ❞**

Driving the PecanPi is Volumio, a popular interface amongst audiophiles. It can accept a number of different services such as Spotify or SoundCloud. In our tests it spotted our local Plex DLNA server immediately and we were playing music without delay. Such is the dedication to pure sound that Bluetooth and wireless LAN are unavailable because the radio interference is unwanted. The PecanPi demands a wired Ethernet connection, and nothing else.

But what a sound. Even with fairly average speakers, *The Dark Side Of The Moon* encoded with FLAC gave amazing detail with a depth and warmth we'd never heard before. This will not be a disappointing product to those who care deeply about how their music sounds. ◾

Verdict

If you can stomach the price, Orchard Audio's humble black Pi-containing box could change the way you think about sound and music.

9 /10

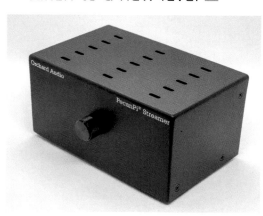
▲ A small black box with an audio monster inside

HiFiBerry DAC2 Pro **& HD**

▶ HiFiBerry ▶ **hifiberry.com** ▶ Pro: £35 / $42; HD: £90 / $108

These two DACs from HiFiBerry are a perfect blend of quality and affordability for premium audio. **PJ Evans** turns it up

SPECS

DAC:
Dedicated
192kHz/24-bit
high-quality
Burr-Brown

SIGNAL-TO-NOISE RATIO:
112 dB

SAMPLE RATES:
44.1-192kHz

COMPATIBILITY:
All 40-pin GPIO
Raspberry Pi
models

▼ The Pro model features an amplified headphone socket and several useful headers

HiFiBerry was one of the first firms on the scene with a range of DACs, (digital-to-analogue converters) that take Raspberry Pi's modest audio capabilities and puts them right up there with the audiophile best. In Raspberry Pi terms, a DAC is an add-on (HAT) that provides much higher quality audio output than you would normally expect.

A rule of thumb is that the more you pay for these boards, the higher sound quality can be achieved (although the law of diminishing returns most definitely applies). HiFiBerry is offering two products, the DAC2 HD and Pro, that sit in sensible places on the spectrum between the basic sound and the ultimate hi-fi.

On the more affordable end is the DAC2 Pro. This features a dedicated 192kHz/24-bit DAC, low-jitter clocks, and low-noise voltage regulators, all with the purpose of producing the best sound possible at that price point. It also features a headphone amplifier for convenience.

A lot more punch
If you're looking for something a little more special and suitable for professional use, the DAC2 HD packs a lot more punch by separating out many of the Pro's components into discrete parts, allowing HiFiBerry to source the best quality in all cases.

> ❝ If you're looking to build a whole-home audio system or something for studio work, these boards are fine choices ❞

Thoughtfully, HiFiBerry offers a dedicated operating system that makes installation as simple as connecting the DAC and powering up. We instantly had features such as Bluetooth and Apple AirPlay with zero effort. In both cases, the sound was impressive and rich. If you're looking to build a whole-home audio system or something for studio work, these boards are fine choices. ⓜ

Verdict

Another impressive bit of hardware design from HiFiBerry, offering both choice and quality at sensible prices. Combined with HiFiBerry OS, these boards are ideal for home audio projects.

8/10

8BitDo Pro 2 **Controller**

SPECS

POWER:
1000 mAh lithium-ion battery or 2 × AA batteries

CONNECTIVITY:
USB-C, Bluetooth 4.0

DIMENSIONS:
153.6 × 100.6 × 64.5 mm, 228 g

▶ 8BitDo   ▶ **magpi.cc/pro2**   ▶ £40 / $50

Professional-grade video game controller that works a treat with Raspberry Pi. **Lucy Hattersley** flexes her thumbs

▼ The Pro 2 controller features all the buttons from a modern console gamepad

8BitDo is a company that's been making a name for itself in the retro gaming sphere, supplying quality game controllers and conversion kits at a good price.

The firm recently sent us a box of interesting things to look at, and we decided to start here, with the Pro 2 controller.

Reminiscent of a PlayStation DualShock, the Pro 2 controller has two analogue sticks, a D-pad, four buttons, four triggers, two Pro-level back

buttons, Select, Start, Star, Heart, and Profile button. It's certainly not short of a button or two.

If all that wasn't enough, there is an 'SADX' Mode switch underneath that swaps between four different modes: Switch, Apple, Android, and Windows.

It comes with a long USB-C cable and 1000 mAh lithium-ion battery with "20 hours of battery life."

In terms of value, £40 is not particularly low-cost in the world of Raspberry Pi, but it is good value when stacked up against its immediate rivals: a Sony DualShock will cost you £50 and an Xbox One Controller starts at £55 (without a rechargeable battery). So, this is cheaper than either. But is it better?

Setup was a breeze. We used the Windows (X) setting on the back and started with a direct USB-C connection. We then held down the Pair button and synced it up with 'Add Device' in the Bluetooth settings in Raspberry Pi OS.

Support in Raspberry Pi OS is game-dependent, although we had a blast in Super Tux Kart and Doom.

We moved onto classic games with Batocera.linux (**batocera.org**) which is a new retro gaming distribution that we'll be talking more about in future. Setup was even easier there, requiring us only to plug in via the USB-C and hold the Pair button. RetroPie was equally easy to set up, mapping the buttons on the controller during the setup process.

For a more modern experience, we tested it out with Xbox Cloud gaming (**xbox.com/play**). This enabled us to use all the analogue sticks and

❝ It integrates neatly with Raspberry Pi ❞

triggers with some of the latest 3D masterpieces. Again, we had no problems.

Button combinations can be mapped to the two Pro buttons on the rear; sadly, the software to control them is only available for Windows, macOS, Android, and iOS. It's a shame you can't do the setup via a web or Linux app.

There are a few quirks. You switch off the controller by holding down the Start button for three seconds, and you might find using Windows (X) not immediately obvious over the other settings. But really there's nothing here that a read of the supplied instruction manual won't clear up.

Holding its own

Build quality of the Pro 2 controller is superb. It's easy to grip and buttons have a nice responsive click with no sponginess. The analogue sticks are weighted well and spring cleanly back to the centre. It's certainly a step above the usual fare for a third-party controller.

The Pro 2 integrates neatly with Raspberry Pi, and the Mode Switch means you can quickly transfer it to any other consoles or computers that you might be using.

We really have no hesitation in recommending this one. ◪

▶ Two Pro buttons, found underneath the gamepad, can be mapped to button combos using separate (non-Linux) software

Verdict

A fantastic controller for a good price that works across a range of Raspberry Pi games, apps, and distributions. Easy to set up and use. Shame the Ultimate Software isn't available in Linux, though.

9/10

CrowPi **L**

▶ Elecrow ▶ **magpi.cc/crowpil** ▶ From £169 / $203

SPECS

DIMENSIONS:
Size: 291 × 190 × 46 mm (L×W×H); Weight 1.1 kg

INPUT/ OUTPUT:
11.6-inch 1366×768 IPS screen; 2-megapixel camera with microphone; 3.5 mm headphone jack; USB keyboard; touchpad; stereo speakers

POWER:
5000 mAh battery (approximately three hours of charge time/ use); DC 12 V 2 A adapter; USB-C interface

▼ Raspberry Pi 4 is fitted magnetically inside the laptop via a clever design

Elecrow laptop turns Raspberry Pi 4 into a battery-powered laptop and electronics learning kit. By **Lucy Hattersley**

CrowPi L is a new laptop from Elecrow. It builds on top of the previous CrowPi2 build that we tested in *The MagPi* issue #97 (**magpi.cc/97**).

It's billed as "a lite version born out of CrowPi2" that reduces the size of the laptop and adds a battery.

The electronics kit is now bundled alongside the laptop instead of fitted beneath the keyboard.

You will need to source your own Raspberry Pi 4. Starting from $203 without the Crowtail kit is good value; even $250 for the CrowPi L and Crowtail electronics kit works out cheaper than its predecessor. Factor in around $50 for shipping.

Design matters

The design is a step forward. The white clamshell case features an 11.6-inch screen, chiclet-style keyboard, and a small touchpad repositioned in the top-right. Included is a 2.4GHz wireless mouse. Poor touchpads plague Raspberry Pi laptops; CrowPi L solves this problem by pushing the touchpad out of the way.

The internal design is clever: you attach four magnets to Raspberry Pi and HDMI expansion board and a 2-in-1 TF card adapter (an A/B switch on this board enables you to swap between two operating systems).

Two ribbon cables connect to Raspberry Pi's USB port on Raspberry Pi and bridge between the HDMI expansion board and CrowPi L's motherboard.

Be aware that we needed a Torx T5 screwdriver to attach the magnets.

Removing a separate panel with a Torx T6 screwdriver reveals the 5000 mAh battery that provides CrowPi L with approximately three hours of runtime.

In the box was a 32GB card with CrowPi's custom operating system based on Raspberry Pi OS (Debian Buster). We also tested out Raspberry Pi OS on the second drive (and via USB boot using an M.2 drive; and Batocera.linux for retro gaming).

Expansion

The GPIO pins are broken out via a smaller 1.2mm pitch 40-pin socket. If you want to use regular HAT hardware with CrowPi L, you'll need the 2.54mm CrowPi L GPIO Breakout board, available from Elecrow for $2 (**magpi.cc/crowpilbreakout**).

Ethernet and USB ports from Raspberry Pi sit on the left; to the right side is a USB-C charging socket, 3.5mm audio minijack connection, full-size HDMI connector, and the smaller 1.27mm pitch GPIO connection (that connects to the Crow Pi L Base Shield with its 20 JST connectors for quick electronics prototyping).

CrowPi L OS

We used the spare microSD port to test Raspberry Pi OS. It runs fine, although we did lose access to the battery charge menu item. Oddly, our screen

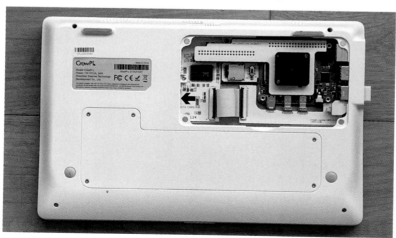

The CrowPi L Base Shield is connected for electronic prototyping

displayed a resolution of 1920×1080 with the stock Raspberry Pi OS, and a look back at config.txt in CrowPi L OS revealed custom timings to set the resolution to 1912×1079.

CrowPi assures us that we have the 1366×768 screen as supplied with all models. We found a resolution of 1280×720 worked best with Raspberry Pi OS.

On the whole, CrowPi L is a nice piece of kit. We're typing up this review on it. It's chunky: measuring 4.5 cm (1¾-inch) at the rear and tapering down to 2 cm at the front and, with Raspberry Pi 4 inside it, weighed it in at 1172 g (2.58 lb). Two speakers offer passable sound and a 2MP webcam worked out of the box. We think it'd be perfectly possible to do a day's work on CrowPi L.

> ❝ We think it'd be perfectly possible to do a day's work on CrowPi L ❞

As with the CrowPi 2, the Crowtail Starter Kit elevates this laptop. The electronics kit comes with 22 modules: LCD, micro-speed motor, 9 g servo, battery pack, and a button, buzzer, and sensors, plus an infrared remote control. There's everything you need here to create a vast range of different builds, and a manual walks you through 21 builds, from Hello World to a remote-control door.

The CrowPi L laptop features a great keyboard, small screen, and tiny touchpad

Verdict

CrowPi L is fantastic value and it delivers a good laptop and great electronics learning experience at a superb price. Recommended!

9 /10

10 Amazing:

Upcycling projects

Take your old tech and make something really cool with it

Upcycling is a proud tradition in making circles and, thanks to Raspberry Pi being so tiny, it's very easy to upcycle old tech... or even just antiques. You could even make it your #MonthOfMaking project if you so wish. **M**

▲ Alexa Ruxpin

Smart teddy

Teddy Ruxpin used to tell stories via cassettes. Now, it will answer all your burning questions using Alexa.

magpi.cc/alexaruxpin

◄ iPod Spotify player

Old school streaming

The iPod is a legendary piece of old tech that, unfortunately, can't quite keep up with the modern streaming landscape. Unless, of course, you install a Raspberry Pi Zero W inside it and connect it to Spotify.

magpi.cc/ipodspotify

▲ Game Boy Zero

Improved handheld

Got an old Game Boy lying around? You could play Tetris with it, or completely gut it, drill into it, and reassemble it as a retro gaming powerhouse with Raspberry Pi.

magpi.cc/gbzero

► Desktop mirror

Retro cool

Reminiscent of speculative tech that you might see in 1960s television, a broken laptop and a pine frame were cobbled together to create this desktop computer that could easily hide on a vanity.

magpi.cc/desktopmirror

▲ Spotify tapes

Streaming cassettes

What if you could put a special tape in an old tape player and turn it into a Spotify media server? Well, in this case, you'd have to put a Raspberry Pi inside first.

magpi.cc/spotifytape

◄ Magic Dresser

Smart antique

Some might say it's sacrilege to modify an antique dresser's mirror to be a smart mirror. We think it doesn't matter what you do with your own stuff.

magpi.cc/magicdresser

▼ Digital dash

Upcycling automobiles

While we're not 100% sure how legal this is, we're very sure that you should take caution if you plan to replicate this very cool upgrade to your car's dashboard.

magpi.cc/digitaldash

▲ 1975 Hitachi Pi Info-TV

Alternate Pi-niverse

This portable television gets a new lease of life in a universe where it was actually powered by a Raspberry Pi.

magpi.cc/infotv

◄ Vintage film camera

35 mm to 35K

Using an old camera housing is a very cool way to make use of a case that it is already fit for purpose, along with some proper lenses, if you can get them to work with a Camera Module.

magpi.cc/vintagecamera

▲ Raspberry Pi boom box

Portable dance party

Instead of streaming Spotify, this upcycled boom box is a portable audio media server using Volumio, but housed in a stylish 1980s facade.

magpi.cc/piboombox

10 Amazing:

Robot projects

Create an automaton with Raspberry Pi

Robotics and Raspberry Pi have gone hand in hand for years, with projects, kits, racing tournaments, and even battles coming from the marriage of microcomputers and mechanics. There's never been a better time to try and build one yourself, so here are ten amazing projects you can try! **M**

▲ Yuri 3

Rover doppelganger

This six-wheeled robot takes inspiration from the ExoMars rover, a future rover planned to be sent to Mars at the end of 2022. This version stays on Earth though.

magpi.cc/82

▶ Reachy

Torso robot

This amazing-looking robot is very interactive and can play games and talk with its owners. It's also all powered by a Raspberry Pi 4.

magpi.cc/reachy

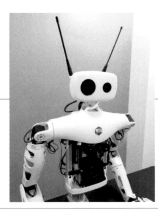

▲ DoodleBorg

Giant towing robot

From the incredible folk at PiBorg, this monstrous robot is powerful enough to tow caravans – as you may have guessed from the trailer hitch on top.

magpi.cc/doodleborg

▲ Build a robot buggy

Your first robot

A great project for building your very first robot, created by the Raspberry Pi Foundation. It's simple and easy to follow.

magpi.cc/robobuggy

▲ C-Turtle

Expendable mine sweeper

This cardboard-based robot is cheap, quick to make, and easily replaceable if it has to take a trip to circuit heaven after completing its explosive job.

magpi.cc/cturtle

▲ Ping Pong Pursuit

Tidy up time

This robot uses a lot of 3D printed parts and a camera to clean up ping pong balls in the 37signals office in Chicago.

magpi.cc/pingpongbot

▶ PARSLEE

Earth rover

This happy robot was made by Dr Jamie Molaro, a NASA scientist, and it includes a seismometer to record earthquakes and other seismic activity.

magpi.cc/83

▲ PiMowBot

Automated green keeping

Take the concept of a Roomba robot vacuum cleaner and add some powerful spinning blades to its underside, and you get PiMowBot – although there's a lot more to it than that.

magpi.cc/pimowbot

▶ Project Zed

Intimidating citizen encouragement

This upcycled robotic wonder was built to help aid people when human contact might not be a good idea. It uses machine learning to give it a bit of a personality.

magpi.cc/projectzed

▶ Build a line-following robot

Simple robot project

This project from the Raspberry Pi Foundation is a great follow-up to the 'build a robot buggy' project, which lets you add some automation to the buggy.

magpi.cc/linefollow

10 Amazing:

Raspberry Pi Pico projects

Make a cool and small project with a Pico

Last month we revealed **Raspberry Pi Pico W – Pico but with wireless LAN!** There are loads of new ways to use Pico, and some projects you can make better with a Pico W. Here are just ten incredible examples of Pico projects to help inspire you. Ⅲ

▲ Cyber glasses

Futuristic wearable

These 3D-printed glasses are more of a base for your own glasses ideas. However, they look great as they are with an LED ring focuser over one eye.

magpi.cc/cyberglasses

▶ Commodore 64 emulator

Retro computing

A C64 expansion board made from a Pico which actually helps the old computer work a bit better than it used to.

magpi.cc/c64pico

▲ Plants with personality

Mood monstera

This plant project makes use of a Pico W's wireless features to text you when it needs water, or generally when it feels like it.

magpi.cc/personalityplant

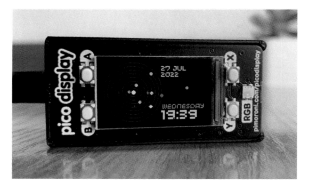

▲ Solar System Display

Space clock

You can use this great project to both tell the time, and also see the orientation of the planets at specific dates and times!

magpi.cc/picosolar

▲ Guitar games controller

Heroic guitar prowess

From the ashes of a broken 3D-printed guitar comes a controller from our friends at HackSpace Magazine to work with a game from our other friends at Wireframe.

magpi.cc/guitarcontroller

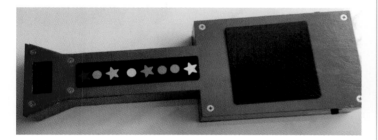

▲ Trill Guitar

Musically inclined

Using some special touch sensors and a custom-made body, you too can create this MIDI guitar.

magpi.cc/trillguitar

◄ Spooky activator

Light it up

This platform allows you to light up whatever is placed on it, and is very simple to make!

magpi.cc/spookactivate

▲ Run a web server on Pico W

Network control

A great example of how to use Pico W from Alasdair Allan of Raspberry Pi. It's incredibly practical too for remote control projects.

magpi.cc/picoserver

◄ Stream deck

What's up, gamers

This specific project uses a Keybow 2040, built on the same chip as Pico. However, you can use the code in the same way with some custom keys added to Pico.

magpi.cc/111

▼ Upgraded Burgerbot

Wireless automated sandwich

Kevin McAleer's excellent Burgerbot is now able to be controlled wirelessly using a Pico W! The whole Pico-based robot build itself is also very impressive.

magpi.cc/burgerbotw

10 Amazing:

Gaming Accessories

Make your Raspberry Pi a lean, mean, gaming machine

We know lots of folks that use a Raspberry Pi to play retro games and homebrew. It's an easier way to hook it up to your TV after all. While you can just bodge together a gaming Raspberry Pi, there are many things that will help make it just that little bit better. M

▲ 8BitDo Pro 2

Retro controller

With a classic retro design updated with modern conveniences (such as Bluetooth), the Pro 2 controller is one of the best ways to experience old games. It works on Raspberry Pi, PC, Switch, and more!

8bitdo.com | £42 / $50

▲ NESPi 4 Case

Retro camouflage

This cheeky case will make your Raspberry Pi seem right at home among other consoles – there's also a removable cartridge for extra storage.

magpi.cc/nespi4 | £28 / $34

▶ GPi Case

Handheld gaming

This familiar case is a genuinely great way to have a bit of retro gaming in your (big) pocket. It even has extra buttons so you can play more modern games as well.

magpi.cc/gpi | £60 / $72

▼ Massive arcade button

For big hits

This huge button is 100 mm across, and comes with an LED as well. It's perfect for very specific arcade and gaming builds, especially if you need to smack a big button very fast.

magpi.cc/massivebutton | £8 / $10

▲ Picade X HAT USB-C

Ultimate arcade board

Created for the amazing Picade, the X HAT is the absolute perfect accessory for turning a Raspberry Pi into an arcade machine. Just build the cabinet, add the buttons, and you're good to go.

magpi.cc/xhat | £16 / $19

▶ Sanwa 8-Way Joystick

Premium movement

Sanwa parts are the gold standard for arcade controllers, and most arcade sticks you get can be easily converted to use one of these sticks. You can even change the gates on them – the corners of the stick.

magpi.cc/sanwajoy | £27 / $33

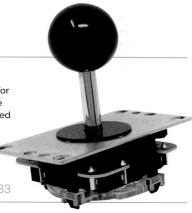

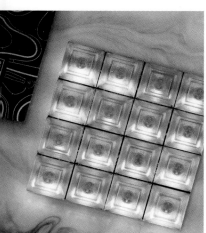

◀ Keybow 2040

Extra hot keys

Need more keys for your gaming? Look no further than the Keybow 2040; 16 mechanical keys based on the same chip as Raspberry Pi Pico. You can program it with custom commands for gaming macros, or stream hot keys.

magpi.cc/keybow | £50 / $60

▲ 8BitDo Arcade Stick

Multi-format stick

Want to get your Street Fighter on like you're at an arcade machine? You can't go far wrong with this arcade stick. It can be wired and wireless, and has many switches to let it work in any way you wish.

8bitdo.com | £78 / $90

▶ RGBerry SMA

Premium RGB HAT

This HAT is designed to connect a Raspberry Pi to an old CRT TV or video monitor much better than the standard analogue video out on Raspberry Pi. See those beautiful scanlines.

magpi.cc/rgberry | £30 / $36

▶ Joy Bonnet

Tiny controller HAT

This funky add-on for Raspberry Pi Zero turns it into a controller that is its own console. It's a bit fiddly, but it's also an excellent party piece, and can even fit in your pocket.

magpi.cc/joybonnet | £15 / $18

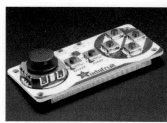

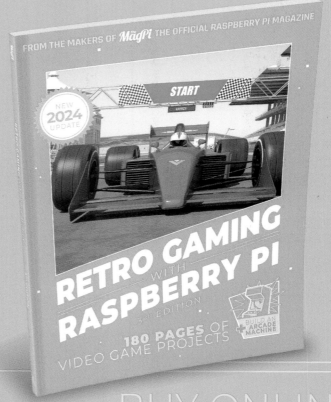